'508

Cable Communications

Other McGraw-Hill Communications Books of Interest

To order or receive additional information on these or any other McGraw-Hill titles, in the United States please call 1-800-822-8158. In other countries, contact your local McGraw-Hill representative.

Cable Communications

Building the Information Infrastructure

Eugene R. Bartlett

McGraw-Hill

New York San Francisco Washington, D.C. Auckland Bogotá
Caracas Lisbon London Madrid Mexico City Milan
Montreal New Delhi San Juan Singapore
Sydney Tokyo Toronto

Library of Congress Cataloging-in-Publication Data
Bartlett, Eugene R.
  Cable communications : building the information infrastructure /
  by Eugene R. Bartlett
      p.   cm.
  Includes index.
  ISBN 0-07-005355-3
  1. Telecommunication systems—United States.   2. Information
superhighway—United States.   I. Title
TK5105.B358   1995
621.382—dc20   95-31568
                                                        CIP

McGraw-Hill

*A Division of The **McGraw·Hill** Companies*

hc 1 2 3 4 5 6 7 8 9   DOC/DOC   9 0 0 9 8 7 6 5

ISBN 0-07-005355-3 (H)

*The sponsoring editor for this book was Steve Chapman, the executive editor was
Robert E. Ostrander, the book editor was Sally Anne Glover, and the production
supervisor was Katherine G. Brown. This book was set in ITC Century Light. It was
composed in Blue Ridge Summit, Pa.*

Printed and bound by R.R. Donnelley & Sons Company, Crawfordsville, Indiana.

McGraw-Hill books are available at special quantity discount to use as premiums and sales
promotions, or for use in corporate training programs. For more information, please write to
the Director of Special Sales, McGraw-Hill, 11 West 19th Street, New York, NY 10011. Or
contact your local bookstore.

Product or brand names used in this book may be trade names or trademarks. Where we
believe that there may be proprietary claims to such trade names or trademarks, the name
has been used with an initial capital or it has been capitalized in the style used by the name
claimant. Regardless of the capitalization used, all such names have been used in an editorial
manner without any intent to convey endorsement of or other affiliation with the name
claimant. Neither the author nor the publisher intends to express any judgment as to the
validity or legal status of any such proprietary claims.

MH95
0053553

Contents

Preface

This book is written for people who are interested in the information highway concept of communications. Actually, an information highway has been around for awhile. The comparison of communications to a system of highways, streets, and roads is a good one. In the beginning, some of the crossroads and connecting streets were rough and needed improvement. As time went by, the network grew and the demands of increased traffic became a problem. In the United States, the information highway was formed by efforts of the government and many of the communications carriers. This effort was put forth to head off any chaos in the communications area. Vice President Al Gore's office has taken the lead necessary to promote the concept and make an orderly and steady approach toward realization of a complete and useful system. Such a telecommunications network is called the national information infrastructure (NII). This network will change as the communications needs change, and it will be continually adapting to the situation as required.

Much has been written about the various sections or phases of the information highway. Unfortunately, many articles are written to impress the readers with the author's importance. Words that are seldom used spring forth in many magazine articles and essays. For example, one article in particular used the word "ubiquitous" many, many times. This word, most will agree, is not often used, and it means widespread, being everywhere—something like pizza. Everyone has one. If the word widespread were used, most readers would not have to interrupt their reading and reach for the dictionary. Another example is the use of the word "paradigm," which means "to show side by side" or a "typical example." Again, simpler language would not detract anything from the writing, and the reader would not have to head for the dictionary. This book is different in that I have attempted to write clearly and present the so-called unvarnished truth. Where possible, I have avoided acronyms because the telecommunications industry has enough of them already. However, a glossary of acronyms and definitions at the end of the book should be helpful for reading further telecommunications material.

This book consists of six chapters that describe and illustrate the various networks and systems that make the so-called road-map structure of the information highway. Chapter 1 covers the concept of the information highway. Chapter 2 explains the various types of wire, cable, and telecommunication conduits. The networks that contribute to the information highway are described in chapter 3. Chapter 4 examines the signals and present uses of the systems discussed in chapter 3. Putting the system together with the various bridges and interconnects is studied in chapter 5, and chapter 6 covers the international interconnect for foreign countries' informational networks. Controls and tariffs are discussed in chapter 6.

The technical level of the material is not difficult for most people interested in information telecommunications. The majority of people in the business, legal, or financial professions with a basic knowledge of communications should have no trouble reading this work. For those with more technical interests, topics of a higher technical level are covered in the appendices. Since this subject is so large and so much has been written, only the current and present issues appear in this book. The last appendix (appendix O) is a more in-depth study of the Synchronous Optical Network (SONET). It is reprinted with permission from the Hewlett Packard Company's CERJAC Division, manufacturers of precision telecommunications instruments. Many other types of digital communications systems use different data-packeting techniques, with headers containing necessary addressing, identification, and control information. But all systems have to do this one way or another. A discussion of SONET technology is included in this book because SONET will be one of the main superhighway toll roads on the information highway.

Acknowledgments

First and foremost, I want to thank my wife, Mona, for her encouragement and help in preparing the manuscript. I also want to thank my son-in-law, Daniel Begeman, for his help in preparing the graphics, and my son, James Bartlett, for his advice and help with the word processing.

A special thanks to Mr. Peter Harper, Marcom Manager of CERJAC, a subsidiary of Hewlett Packard, for obtaining permission to include the SONET Networks and Testing Techniques as appendix O.

Defining the Information Highway

A highway is often envisioned as a standard vehicular roadway going in two directions, to and from, with access and exit ramps for entering and leaving traffic. Traffic flow in a general, limited-access highway is typically asynchronous in that each vehicle essentially travels at its own speed. Often, the lower and upper speed limits attempt to place some restrictions on the traffic flow. Unfortunately, collisions do occur, and results are often catastrophic. Other types of highways attempt to control traffic by using traffic signal lights. Such traffic signals, the usual "stop" and "go" lights, cause traffic to group together and bunch up, which is a sort of synchronizing control.

The one communication system in the world today that resembles a true highway is the telephone system. Another type of communication system is the bidirectional computer network called a LAN (local area network), which usually covers a building or campus complex. Expanded networks of this type are labeled WANs (wide area networks), and such networks that envelop a municipality are called MANs (municipal area networks). Some cable television systems also have the capacity to operate in a bidirectional manner.

Since the federal government is now in the process of deregulating the communications industry, it seems attractive to merge the various companies in the communications business into enterprises that combine their plants and systems into an intercommunications network. Of course, the following questions should be asked: Is there a ready market for such a system, and what are the users willing to pay for it? The fact remains that any business venture has to stand on a sound economic footing.

1.1 Users of the Information Highway

The present users of bidirectional information networks most likely will continue to use the systems, and their needs should expand if their businesses dictate it. If the use of an information system provides fast and accurate customer service, it should place the company in an advantageous and competitive climate. This in turn makes the system user-efficient and cost-effective. Of course, more business should result, and the using company should prosper.

1.1.1 Commercial uses and applications

Many companies in the United States use and require information-communications services in various degrees. Such companies fall into several major categories, from simple telephone/computer-modem sets to high-speed digital data and or video/audio teleconferencing requirements. The geographic distance between users plays an important role in the type of communication system that will meet the users' needs, and this affects resulting costs.

1.1.1.1 Financial institutions. The financial community is indeed a large user of bidirectional communications systems. Banks, many of which use satellite links, also use telephone-system modems to carry financial digital data between bank branches, home offices, and main accounting offices. Credit-card companies also transfer credit, sales, and accounting information on bidirectional communications systems. Examination of credit histories, real-estate property evaluations, and all sorts of financial data are transferred among offices all over the country.

Any information-highway communications system should consider and investigate thoroughly the present and future needs of the financial community. Data rates of digital computer data and any present and future teleconferencing requirements are the types of information needed for planning such a system. Locations of future offices or areas where connections might be required should also be made clear during the planning stage.

1.1.1.2 The medical community uses. The medical community appears to be starting to enter the communications arena. Most of us know that during an appointment with our physician, notes are made on paper. However, the various diagnostic imaging methods employ various scanning techniques of either X-ray, NMR (nuclear magnetic resonance), cardiac echograms, or catherization systems. These yield digital data that is often processed by computer software to enhance the image presented to the physicians. Storage of this type of medical information can be made by conventional photographic film, videotape, or floppy disk, and transferring such data over the information highway will most likely be required. Any video techniques will require much more bandwidth than some other types of data-transmission methods. Such video pictures could provide graphic information on a patient on vacation, where medical records are at a distant location.

It is expected that the physician will soon have a pocket computer available that is loaded with the next patient's medical records. During the patient examination, the physician can look at the records, make any additions and notes, and at the conclusion of the examination, dump the information into the local office data bank. This would be followed by a reloading procedure for the next patient. A method such as this means that patient records will be placed on the local office file in digital form and be available for transfer to other areas via the information highway. The reverse procedure would supply patient data from a former location to the present one where the patient resides. Some major teaching hospitals already have some communication systems in place, but such systems are confined to local urban areas.

Expansion of the medical use of the information highway and its benefits to the medical community should be thoroughly investigated. Health-care costs could be limited by reducing the time physicians have to spend manually searching records and examining the myriad of drugs for results and side effects. The benefit to patients could be even larger, particularly if patients are traveling or away from personal physicians. This is an area where a selling job needs to be done because the benefits of the information highway to the medical community can be enormous.

1.1.1.3 Industry and manufacturers. Industry seems to be always having financial problems, so the cost of producing products is extremely important. Many manufacturing companies in the United States have in the past used what is known as numerical-controlled machinery. Such machinery has evolved into so-called robotic machines that are microprocessor controlled. With the advent of the personal computer (PC) and parallel and serial inputs, the microprocessor-controlled machine could communicate with the personal computer. Control information from the monitoring personal computer could be sent to the various machines on the manufacturing floor. When software was developed for such applications, the term CAM (computer-aided manufacturing) came into being. Since the design process could also use PCs, software for CAD (computer-aided design) was developed. The next natural step was to develop the CAM process as an add-on to the CAD process. Thus the combined system became known as CAD/CAM. Final testing also used a sort of CAD/CAM method, which became known as ATE (automated testing and evaluation). Both CAD/CAM and ATE techniques employed digital data that needed to be sent to the controlling computer for comparison against the specified values. Corrective controlling data was sent back to the machine's local microprocessor. Thus the manufacturing system became a controlled loop.

In general, many manufacturing plants operate in what is known as a very noisy electrical environment. The electrical noise generated tends to interfere with the transmission and reception of data. In the presence of noise, digital data is definitely more robust than analog data methods. However, proper shielding, error checking, and clear electrical power are some methods used to preserve the data and prevent errors.

Many manufacturing plants have installed various cable systems to handle the flow of information. The first LANs (local area networks) were installed in factories to provide the flow of control information. Some factories employed mechanical communication lines in the form of pneumatic or hydraulic control lines. Changes in pressure corresponded to the control signals, and many lines had to be installed in order to provide a closed-loop condition between the machine and controlling element. This method was immune to electrical noise, but it was extremely slow in communicating information.

Since many companies that manufacture products in the United States are very large, data communications between the home office and the various local factories became necessary. For low data rates, the typical computer telephone-modem technique was sufficient, but some companies have now opted for a satellite communication system with data packet switching. This method allows very high speed data transfer in packets. Theoretically, various manufacturing facilities could be controlled from a remote location through such a high-speed data communication medium.

Some industrial-type LANs employ a bidirectional coaxial cable system that is similar to some of the two-way cable television systems in operation today. Such a system is used in the automobile manufacturing industry, where digital data is transferred on RF (radio frequency) carriers employing modulation/demodulation methods. One such system uses the designation of MAP/TOP (manufacturing automation protocol/technical and office protocols).

To bridge the gap between various manufacturing facilities and their home offices, the information highway could provide effective data communications, provided, of course, that it is cost-effective. Several of the automobile-manufacturing companies presently use a satellite system of bridging the gap, so to speak. Sales and credit information, not manufacturing information, make up the data traffic.

1.1.1.4 Utility company uses. The utility companies also have major communications needs. Meter reading is still often performed by a person reading the meter indicators with the aid of a hand-held data-entry computer or a bar-code device. Later, this information is transferred to the local data storage and computation center for usage analysis and customer billing. There have been cases where electric, gas, and/or water meters with digital data outputs have been used in conjunction with either telephone or cable TV methods of data communications to the local office. The myriad of older meters that are still performing adequately and the high cost of changing out to an automatic system cause many utility companies to use personnel as meter readers. The day will come when more and more utility companies will upgrade to an automatic digital system, and the information highway will be able to provide the data communications link.

Utility companies use plant and system controls that are connected by a variety of data communications methods. The electric and gas generating/transmission companies have systems in place to connect switching and distribution stations together.

The electric utilities have high-voltage transmission structures placed along rights-of-way paths, which connect to switching and regulating stations of points of electric service distribution. These stations are often found in remote areas that are often unattended by personnel for days. Monitoring of intrusion, environmental status, and operating parameters through online measuring instruments are the types of information required by the utility operators. This data is usually transmitted to a central control station by digital techniques. The communication link is either through telephone modems, a radio control link, or both, with one method the primary method and the other a backup. Again, the information highway could provide the monitoring functions and downstream control instructions and commands to the remote switching and regulating stations.

In a similar fashion, gas pipelines are buried in rights-of-way where gas is transmitted to distribution stations for distribution to local gas companies. At various stations along the way where distribution, pumping, or whatever take place, monitoring of pressure, temperature, leaks, intrusion, and building environmental parameters are the types of usual measurements. Again, such data usually appears in digital form for transmission to the office on control stations along some communication medium. Some sophisticated utility companies, both gas and electric, are using or have planned fiber-optic data communications cables along the rights-of-way routes. Electrical interference from high-voltage systems do not affect the optical data. Hence it is immune from electrical noise ingress.

Some states have a system of canals or rivers flowing through the area. Since water management is so important to the environment, control of water dams and gates is extremely important. Control and monitoring data generated at many remote, unattended dam sites are transmitted through telephone company lines using modems and/or a radio control system. The information-highway-planning people should consider such municipalities' water-control communications needs.

1.1.2 Government uses and needs

Some of the largest users of bidirectional communications are the federal, state, and municipal governments. Through its own supervised telephone network, the federal government is tied into the operating telephone systems in the United States. The Bell System was contracted to provide data communications for the Departments of Defense and Transportation (FAA). Equipment was installed, maintained, and managed under such contractual arrangements.

1.1.2.1 Defense and security. The federal government has in the past used the telephone industry's long-haul wide-bandwidth expertise in constructing a microwave facility to service remote radar stations and provide monitoring information to area military installations. The radar sites were part of what was known as NORAD (North American Air Defense) system. The telephone system in the United States was definitely there when the government needed it. USAF radar sites of the PAVE PAWS system most likely use portions of the telephone system to distribute radar data and information to the Department of Defense. Again, the information highway could be a benefit to such radar installations by providing high-speed data channels to the users of the radar information.

1.1.2.2 Law enforcement. The United States Department of Justice in the law enforcement section is probably a large user of available data communications systems. Video information, fingerprint information, and other types of personal identification information needed by law enforcement officials are transmitted and received today by the telephone system and/or by satellite communication systems. Vehicle registration information is presently transferred through commercially available communications systems, principally the telephone system. Since the federal government as well as the state and local governments are cracking down on the drug problem, increased data communications are needed. The information highway could provide the law enforcement communications needs for a long time to come.

1.1.2.3 Post office. Most of us are aware of the use of computer systems by the post office. Obviously, the local office computer data is transported most likely by telephone modem to area post offices. Regional offices can be connected by telephone company systems and/or by a satellite radio system to area accounting and control offices. A well-managed, high-speed wide-bandwidth bidirectional information highway could provide proper communications for the United States postal system. If such communications facilities became available, the postal system could then look to upgrading to bar-coded stamps and addresses for automatic mail sorting and routing.

1.1.2.4 Department of Transportation/FAA control information. The Department of Transportation of the federal government is a heavy user of data, voice, and radar data communications. This is mainly from the Federal Aviation Administration (FAA) part of the Transportation Department.

Again, available commercial systems, mainly the telephone system, are the communication carriers. Most of us remember the computer glitch that caused the FAA system to fail, leaving air traffic in a quandary a few short years ago. The FAA is presently putting together an upgraded system with more fail-safe features. The information highway should have the ability to provide the necessary communication links to encompass the FAA needs and for the Department of Transportation in general.

1.1.3 Home and private uses

In today's world, the average domestic consumer pays the telephone bill and the cable television bill. In many homes the telephone bill ranges from a simple service with a low volume of toll calls of 30 dollars to oftentimes about 200 dollars. Cable television service charges range from about 17 dollars to about 60 dollars for heavy users of premium (pay) services. How much more will the domestic customer be willing to pay is one of the most-often-asked questions. Clearly, the benefit of the information highway to the domestic consumer will have to be plainly presented in order to entice users. Services to the consumer that could be connected to the information highway should be thoroughly investigated.

1.1.3.1 Entertainment. The telephone as we know it is often referred to as POTS (plain old telephone service), and it should be carried with ease on the information highway. Since this service will only occupy channel space when it is used, other services normally piggybacked on the telephone system will be handled separately on the information highway. Consumer equipment will be developed to control voice two-way (telephone) service on the information highway and on the home computer information-interface devices.

1.1.3.2 Personal computer information exchange. Most home computer systems today are equipped with telephone modems. These modems can be commanded by the telephone ringing. If nobody is home, the computer will be activated to receive messages of digital information from a sending computer system. When the sending computer is finished, it hangs up (a disconnect), which now tells the receiving computer to hang up. The result is a computerized answering system where the receiving computer stores the message in its memory. When desired, home users can turn on their computers and read the messages on the screen. This is part of the e-mail concept. Such users are already paying this charge, which now appears on the telephone bill. If the charges for the information highway are no more than the present telephone bill, then current users will continue. If indeed the cost is reduced or more enhanced services are available, new customers will be added.

1.1.3.3 Voice/telephone uses. To the home consumer, entertainment is very important. This comes mainly today in the form of cable television service. As most of us know, this service provides network television, independent television stations, superstations, pay channels, religious channels, sports channels, and a group of informational and shopping services. How much more the consumer is willing to buy is a good question and should be considered. Many have thought or are thinking video-on-demand will be the way to go. This service requires the consumer to order the desired video through the set-top control interface, which is then delivered through the same interface in the desired time slot. Again, enhanced services and added cost result. It is not enough to just make the service possible. It has to be desired by the consumer who has the money and is willing to pay for it.

Another form of video entertainment is playing video games. Many users of home computers often have a few of the common video games available on diskettes. Some parents of children might think that if better video game software were available, the children would possibly stay home and out of trouble by avoiding the many video-game parlors that have sprung up in many neighborhoods. This might be true, and an information highway could provide ordered software that could be received through the interface and recorded on hard disk or written on diskette. The video game store could service its customers through the information highway. The transferring of software from person to person is common today and causes a problem with copyright laws. How such interaction will be affected by an active, two-way information highway is a searching question. Again, the associated costs for such services will dictate the success of the venture.

One feature of the information highway that many domestic users might be interested in is the home security or alarm system service. Presently, many home security methods use the telephone system either through a dedicated line or a dial-up feature. Some more elegant alarm systems have a radio backup system in case the telephone line is severed. If indeed the information highway had enough bandwidth so video information could be sent to the central monitoring point, a true surveillance alarm system could result. Such surveillance would not happen until possibly a perimeter alarm triggered the video system. Upon receipt of the video information, the entering party could be identified either as proper or as the intrusion of a burglar, and a video recording of the event could be made while authorities are heading to the scene. Also, verbal warnings to intruders could be made to scare them away. An information highway could actually be used by an area surveillance system covering a whole neighborhood or development. Such costs could be shared or paid through a homeowners' association.

1.2   Status of Information Highway in Place Today

There are several parts of an information highway in place today. These are operated as separate systems offering services to subscribing customers. Of course, the largest system is the telephone system. The cable television system is another system of communication in place, as well as local area networks (LANs) operating in office parks and industrial areas. Also, municipal systems such as cable systems servicing municipal offices are often referred to as metropolitan area networks (MANs), and they appear in many major cities in the country. A college campus often has a cable system connecting various buildings with computer information, alarm and intrusion surveillance, and television for education and entertainment. These systems are sometimes referred to as wide area networks (WANs). The use and status of these in-place systems should be considered as parts available to the information highway.

1.2.1   The telephone companies

The telephone system in the United States is one of the most sophisticated and reliable bidirectional communication systems in existence today. The technical aspects of the present-day system will be discussed in the ensuing chapters.

The telephone system developed from the primitive invention of Alexander Graham Bell to a very reliable, high-quality system in the United States. The company philosophy was always customer-oriented, supplying good, reliable services at affordable prices. Everything the company did always seemed to make sense. From the local installer to the area technicians and engineering staff, good people were hired and a lot of time and money were spent on employee training. Those who did not measure up were discharged from the company. High standards for plant, equipment, and methods of doing business were formed early in the development of the company. The Bell system, as it became known, formed its own research facility and its own manufacturing facility. Bell Telephone Laboratories, known simply as Bell Labs, has enjoyed a long and illustrious career as a model for high-quality corporate research. The Western Electric Company, the manufacturing arm for the Bell system, was credited with developing the nearly indestructible black subscriber telephone set. During World War II, the U.S. Army Signal Corp literally adopted the Bell System Specifications for telephone plant construction and maintenance, right down to the olive drab installer trucks and the pole-setting vehicles. The Western Electric Company was one of the main manufacturers of telephone equipment for the Signal Corp during the war. The U.S. government relied heavily on the telephone company to fulfill the communications needs for the war effort. And the telephone company indeed did so.

1.2.1.1 Voice communication. The main concern of the telephone system today is voice communications. There is seldom a home in the United States without at least one basic telephone set. The telephone service is considered by nearly everyone to be almost a necessity of life. Essentially, the local exchange that served a town had an unobtrusive building that resembled the surrounding homes, where all the area twisted-pair cables were connected to a switch-bank frame. Local calls were switched to other local pairs by the switch bank. Toll calls or no-charge calls from one local exchange to another were switched to trunk lines connecting such areas. Essentially, local exchange systems are referred to as hard-wired electromechanical systems. However, the electronic equivalent of the switch bank is currently used in many installations. Some rural areas possibly are still operating with the old electromechanical relay switches. New equipment was always completely compatible with the old types so overall operation was still reliable and of high quality. Most of us recall that touch-tone service and pulse dialing were both operational on the same local system. Most modern telephones used today are electronic. A microchip performs the tone-dialing signals and duplicates the action of the old transformer-type hybrid and, in general, the operation of the telephone set. Plugging in an older telephone set will still work fine. It must be noted that the present-day telephone system is a full duplex system carrying audio information (voice) between connected parties. In this respect it is a true information highway that connects business, commercial, and home subscribers.

1.2.1.2 Computer/modems digital communications. The telephone industry indeed knows its product well. The quality and specifications of their various lines and channels are known. Some telephone company routes have better noise figures than other routes, while some routes have more bandwidth than others. All this information is accurately logged in the company files and records. When many businesses were becoming computerized, the need for computers and data banks to communicate arose. Both the computer industry and the telephone industry took advantage of the situation, and computer communication modems using the telephone system were developed. The semiconductor manufacturers developed the necessary chip sets and semiconductor devices needed to implement this new technique. Many people in the industry remember quite well the universal asynchronous receiver-transmitter semiconductor device termed the U-ART, which is still in use today. Present-day modems have either switch-selectable or software-selectable data rates so the data rate selected can conform to the quality of the telephone line available. In this manner, many businesses are connected by computer modems transferring computer data between locations. The speed of data transfer depends on the quality of the telephone connection. This requirement has been met in many cases by the telephone company's ability to reroute the connection to improved lines needed to accommodate the needs of the customer, and such use of the telephone system to transfer computer data has resulted in cost savings to many customers. Such cost savings by American industry effects the overall economy of the country. Many other countries are doing the same thing with their telephone system, which has resulted in sales of equipment, systems, and techniques by manufacturers in the United States. Also, the reverse is happening with foreign manufacturers selling their communication products in this country.

1.2.1.3 Telephone network interconnects. The telephone network in this country is a conglomeration of various network topologies. The simplest is the local exchange servicing a small group of customers ranging from local industries, municipalities, and homes. Such a local exchange is shown in Figure 1.1.

Service between local customers does not go beyond the local exchange area. A call originated by a local telephone set is acknowledged by the local exchange office and is switched to the called party's line to complete the connection. Once the call is completed, the telephones involved hang up and the lines are released. Calls originating at a local exchange to an area beyond the local exchange's control are routed via trunk lines to a higher-level exchange area, eventually arriving at the lo-

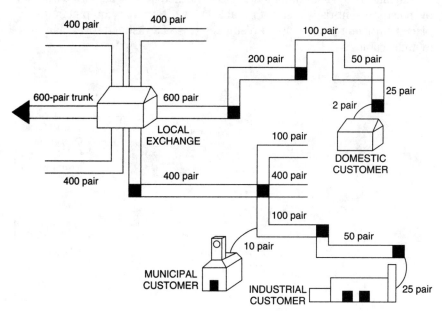

Figure 1.1 Elementary diagram of a local telephone exchange cable plant.

cal exchange of the called party. The connection between the telephone sets of the calling and called party is maintained as long as the off-hook condition continues to exist. The calling (originating) party is charged for the time used for the connection. Since trunk lines are connections between exchanges, the supervision of trunk connections is performed by the higher classified exchange. Figure 1.2 illustrates typical connections between several classes of exchanges. Local exchanges are often connected to toll centers that might be located in a more metropolitan area. These higher classed exchanges are connected to primary area exchanges, often by multichannel coaxial cable and/or by a microwave radio link. Several primary exchanges are connected to a larger sectional exchange that controls message traffic in a section of the country.

The telephone industry uses various types of communications techniques to transfer message traffic in the form of voice and computer data. Microwave multichannel radio links, multichannel coaxial cable, and satellite radio transponder links are the more usual types. Lately fiber-optic techniques are being employed and are gaining widespread acceptance as the communication method of choice. The telephone industry has been placing a large amount of fiber-optic plant along existing routes and rights-of-way. This type of cable is an electrical insulator and hence is not prone to lightning and electrical noise ingress. Also, each fiber can support a large frequency bandwidth, which presently is only limited by the terminal transmitting and receiving equipment. The extremely low signal loss exhibited by most present-day fiber-optic cable allows terminal equipment to be placed sufficiently far apart in distance. This is a distinct advantage over metallic cable.

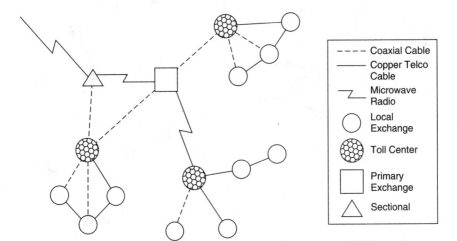

Figure 1.2 Long haul telephone interconnect.

1.2.2 The cable television companies

Cable television has become a large and influential communications industry. It has been estimated that nearly 80 percent of the homes in the United
States could be served with cable television service. Most present-day cable
television companies have expanded their cable plant far beyond the beginning 5- and 12-television-channel systems. The more progressive companies have the capability of transmitting at frequencies up to 1 GHz (1000
MHz) covering nearly 150 downstream television channels. Many companies also have the capability of using some of their cable bandwidth for reverse or upstream service. However, most of the channel bandwidth is
reserved for downstream delivery of video service (ordinary television) to
domestic home subscribers.

1.2.2.1 Video services. As stated earlier, the principal product that cable
television companies offer is essentially video services—normal television
video/audio service. That is the meat and potatoes of the cable television
industry. The NTSC (National Television Study Committee) signal is a complicated signal to transmit. This signal contains pulsed signals required for
synchronization, wideband video signals for the luminescence or brightness, and quadrature double-sideband suppressed carrier for chroma information. Frequency-modulated wideband stereo audio accompanies the
video signal. This uses a 6-MHz-wide standard television channel.

Sources of such video signals originate at the television broadcast stations and are received by cable television operators usually by an off-air receiving antenna tower. Often each station is received by a single antenna
array, and the signal is processed by a single channel processor. Other video
services are received by cable companies via a receive-only satellite ground
station. A multitude of video programming is available via satellite, including the premium or so-called pay TV program suppliers. The central receiving area for a cable television installation is often referred to as the
"head-end," i.e., the beginning or source of the signals offered for service to
subscribers. In many cases a good head-end system might be connected to
another company hub-site and so might act as several head-ends. The communication medium used to connect a head-end to a number of hub-sites
varies from a supertrunk coaxial cable system, to a microwave multichannel
radio system, or a fiber-optic system.

Another source of video programming is the local area studio, often
referred to as local origination. In many cases this was a requirement issued
by the local licensing or franchising authority. The cable television company, often known as the licensee, was required to produce local programming at a local studio consisting of programs of local interest. Some
licensing authorities often require several channels for local use, such as a
local school channel, a municipal channel, and a public information chan-

nel. Many cable television companies regarded such services as a necessary evil. However, many companies rose to the occasion and produced interesting, educational, and informative local programming. Since the only people who received such programming were cable subscribers, it was indeed an inducement to subscribe. Hence what started as a problem ended up becoming an aid in selling the service.

1.2.2.2 The cable television distribution plant. The cable television plant is indeed quite different from the telephone plant. As stated earlier, the telephone system is a channel-by-channel full-duplex system. The cable television system is basically a one-way downstream head-end-to-subscriber system offering wideband multichannel television service. An example of typical cable television network topology is shown in Figure 1.3. Still, many companies have a reverse or upstream capability. The number of upstream channels and the number of downstream channels are not equal. Earlier designed bidirectional cable television systems employed what is termed in the industry as subsplit reverse. Frequency channel allocations for such a system assigned 11 channels in the band of 5.75 to 47.55 MHz. In practice, it was only the 4 to 5 lower channels (T-7 to T-11) covering the band of 5.75–30 MHz. Only ½ of T-11 could be used. This prohibited band of 30–47.55 MHz was reserved for the crossover filter band between the forward (downstream) am-

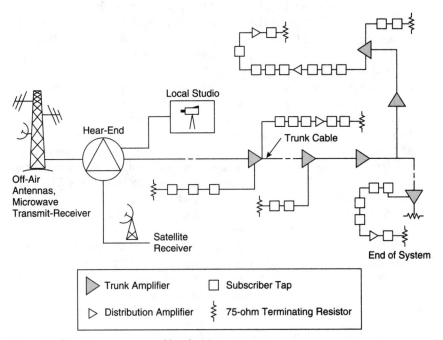

Figure 1.3 Elementary one-way cable television system.

plifier cascade and the reverse (upstream). In short, there were many down-stream channels and only 4½ upstream television channels for a subsplit reverse cable system.

Other bidirectional systems were later designed that allowed for more upstream bandwidth. First was the midsplit, allowing for nearly 18 6-MHz-wide television channels upstream. Such systems with 600-MHz upper bandwidth would have the 18 upstream channels in the 5–112 MHz band and 75 television channels in the downstream band of 150–600 MHz.

A high-split system was developed later that allowed for 61 channels downstream and about 28 channels reverse. This type of system was seldom used by the cable television industry except in highly populated metropolitan areas requiring many reverse channels. In general, most cable systems in business today do not use any reverse or upstream services.

As part of the licensing or franchise agreement, some municipalities required the licensed cable television operator to provide a separate section of cable television plant to be constructed for exclusive use by the municipal government. Such systems often were bidirectional in nature in either a subsplit, midsplit, or high-split single cable system or in some instances a dual cable system with a separate cable system for each direction. In some instances, municipalities carried video services for educational purposes, video services for surveillance of municipal properties, and computerized digital information between various municipal computer sites. Audio channel communications in the form of alarms or telephone-intercom services were carried on such networks. Cable television operators were often surprised by some of the innovative uses many municipalities implemented on their municipal networks. Still, many cable operators who constructed systems for municipalities who did nothing with them regarded the requirement as a waste and the so-called price of obtaining the license to build the cable system.

1.2.2.3 Audio/radio services. A closer look at a cable television operation tells more about how the total downstream bandwidth is used. Early systems carried television programming on the assigned broadcast channels, since a television set connected to the cable system could indeed tune to that channel as if it were from a local antenna. Hence the acronym CATV was given, which stands for community antenna television. Since the FM radio band falls between television channel 6 and 7, it too was carried on a cable system, allowing subscribers to connect their FM stereo receivers to the cable and receive improved FM reception. The space between channels 6 and 7 also contains an amateur radio band and an aircraft communications band, which does not appear on the cable. Cable operators developed 9 more 6-MHz television channels to occupy this space. Unfortunately, current television sets could not tune to those channels, so the cable television industry developed a converter box allowing a television set to view these channels. With the arrival of satellite service, many more sources of television programming were forthcoming, filling the use of these 9 so-called midband channels. The use of these channels also made it very necessary for cable television plant to be maintained very tightly so signals would not leak out and cause interference, particularly in the aircraft emergency communication band. The Federal Communications Commission (FCC), the regulating body for communications in the United States, has specific rules monitoring and reporting procedures that control cable television plant signal leakage.

As upper frequency limits for cable systems increased, the channel capacity also increased. More and more television services are being offered by many cable systems. As more channels were added above the standard channel 13, a more sophisticated converter was developed to allow a normal television set to receive these services. Broadcast television stations in the UHF (channel 14–83) band were converted by the cable television operator to a frequency band that the cable system could handle. The newer converters allowed television sets to view these translated UHF channels. Enter the cable-ready television set.

Various means were developed by the cable industry to withhold or prevent certain channels from being received by a subscriber. These services are known as simply the pay channels or premium channels, which cost extra. Narrowband notch filters (traps) placed in a subscriber drop cable to the house would trap out the premium channel. If a subscriber ordered the service, the trap was simply removed, allowing the subscriber to tune in the channel. Since this system was not regarded as very secure (some subscribers removed their trap), scrambling methods were employed. Such systems remove and/or alter the signal, and if a subscriber ordered the service, the converter could be programmed to receive the key, so to speak, and restore the altered signal to normal. A PROM (programmable only memory) was often used by some converter boxes as the decoding element. A cable technician was required to install the PROM device for the converter to restore the signal.

In later years the converters were made so-called smarter with the addition of addressability. The addressability feature required that the head-end had to transmit the address of each converter every so often to keep it active. Also, the ordered services were sent so the converter could unscramble the ordered premium channels. Various techniques and methods were employed to transmit this data, depending on the manufacturers of the equipment. Cable-ready television sets could tune to the unscrambled cable channel configurations but still needed the converter box to unscramble the premium services. Unfortunately, the cable-ready television, the video cassette recorder (VCR), and the converter could be used as a channel selector, with each having its own remote control. This situation was regarded as extremely user unfriendly.

Addressable converters with built-in unscrambling were used in large, more metropolitan cable systems. Still, pirate converters (illicit) were being sold to circumvent paying for the premium service. Stolen converters were being sold, and theft of service became a problem for cable operators. The telephone industry suffered from a similar situation when tone generators with stolen codes were being sold, which enabled calls to be made without paying for them and/or the wrong party being charged. Some large cable operators experimented with various methods that required the converter to answer back to the addressing computer either through one of the system's reverse channels or the telephone systems. Such a system enabled cable operators to verify the location of the box, that each converter was active, and that the provided service had been ordered. Stolen converters now could be located to some degree, thus improving system security.

1.2.2.4 Digital information capabilities. All of the foregoing operations required digital information to be carried on the cable system, and the cable operators became aware that their cable systems could carry much more than entertainment. It must also be remembered that the primary service provided by cable television operators is entertainment. Hence cable television drops were made mainly in residential areas, not industrial or commercial areas. Therefore many cable systems do not have plants available in industrial-commercial places that would be among the first users of the information highways. Also, many cable television systems are still one-way, i.e., downstream, so in order to become a true player in the information highway, an active reverse system will have to be added. Interchange of digital data between various locations is one of the main uses of the information highway. Such digital data will also include compressed digital video/audio signals for entertainment or instructional television service, along with digital computer data simultaneously.

1.2.3 Other types of cable systems

Industrial needs for communications caused certain companies to form their own systems. When such needs called for more frequency bandwidth than the local telephone company could provide, a company would construct its own system. Commercial power companies, the railroad industry, and large manufacturing plants constructed their own so-called in-house communication facilities. Some such systems became quite sophisticated and large, often encompassing a considerable area of the country.

1.2.3.1 Utility company communication methods. The commercial power utility companies control and operate a large power distribution system to distribute electrical energy to their customers. These systems from company to company are connected to form the national power grid. The communication systems are also interconnected so each company can sell excess electrical energy and also buy energy from the grid. Information on metering the flow of energy is necessary, so the various power companies can get paid or charged for the energy sold or bought. Also, load-switching instructions, testing, measurement, control, and fault-testing information are carried on the communications system. The power industry communications systems often consisted of radio and microwave links. Sometimes a wired system was used. However, as the electrical voltage was increased on the power system, noise became a problem with wired systems. As stated earlier, the power grid essentially covers all of North America and Canada, with the high-voltage lines traveling the rights-of-way from city to city where distribution centers are located. Many power companies are considering or have installed fiber-optic cable along these rights-of-way. Since the glass fiber cable is an insulator and optical transmission is in the micrometer wavelengths, electrical noise is not a problem. Usually, when fiber-optic cable is installed a single cable contains many individual fibers, enabling a huge amount of information bandwidth, far more than the power industry needs. This excess information capacity could certainly be sold to operators of any information interconnect. Thus, many power utilities could find themselves involved with an information highway and the benefits of substantial added revenue.

1.2.3.2 Railroad company communication system. The railroad system in the United States operates along rights-of-way that have been in existence for a long time. From the beginning, the railroads have been in the business of communications by carrying the mail to the telegraph lines that were built along the tracks. The telegraph system in use in this country for many years was developed from the railroad company's telegraph system. The railroads saw the possibility of extra revenue by building a microwave radio system along the long straight and flat western routes. Most of the channels were leased to common communication carriers for telephone traffic and broadcast network television feeds. The railroad operators maintained and operated such systems for many years, and many are still in operation. Broadcast television network feeds switched over to satellite systems when they became available. However, for geostationary satellites, an eclipse of the satellite occurs twice a year for periods of several minutes, causing a system outage. Communications carried on a satellite system had to be temporarily switched to a satellite not involved in an eclipse. Most likely, the signal quality, reliability, and cost factor made satellite communications attractive to common communications carriers over long microwave routes. Still, the railroads have the rights-of-way and a fiber-optic system built along those rights-of-way would certainly make sense and be more reliable and less expensive to maintain than a microwave system.

1.2.3.3 Computer, local, wide, and municipal communities. When computer systems were developed to a point where multitasking and file serving were a normal part of a system requirement, the need for a communications link among computer workstations became necessary. Computer equipment manufacturers were quick to supply hardware and software that facilitated the interconnection. Hence the so-called local area network (LAN) was born. In the beginning, a multiwire cable system from 2 to 24 parallel wires formed the RS-232 interface standard. This standard specifies how digital data is moved from one device to another. This was the standard that also allowed connection to the telephone system through a separate modem device or a built-in modem card. This indeed was a rudimentary but effective method used in data communications equipment. Naturally, higher data rates were required when faster computer equipment was developed.

The cabling that connected equipment in the beginning was twisted-pair wire, which was limited as to speed and distance between devices. Coaxial cable became quite commonly used and supported such well-known LANs as Ethernet. Presently, fiber-optic cable with its wideband benefits makes possible high-speed data transfer rates with a minimum of amplifying or repeater devices. The use of coaxial cable and repeater amplifiers similar in design to those used in cable television systems allowed the area of network coverage to expand in size. These larger networks are known as wide area networks (WANs). Such systems allowed high-speed data to be interchanged in both directions since the system had an outbound link and an inbound link. Figure 1.4 illustrates such a system. Modulated carriers of differing frequencies carry the data to the various computer workstations.

The success of various LANs and expansion sufficient to become a WAN caused many municipalities to construct a metropolitan LAN/WAN system called a metropolitan area network (MAN). Operation of many MAN systems was identical to LAN systems. Municipal governments in some instances require a cable television applicant to include such a network in its license application. Hence, many municipalities were able to have a net-

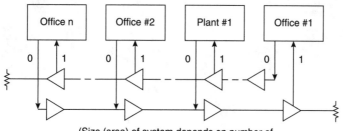

(Size (area) of system depends on number of
input/output data channels, cable loss, and data speed)

Figure 1.4 Example of a local area network.

work installed at little or no cost to the taxpayer. Voice, video, and computer digital data were carried on these full-duplex networks—a true information highway in miniature. All municipal offices, stations, and schools were connected by such a system.

It should be evident by the material covered in this chapter that there is a tremendous benefit in having a system capable of interchanging, in real time, information between locations in various areas of our country. Such information can be voice, pictures, and/or computer digital data.

Economically, it could have a great impact on such areas as health care, law enforcement, commerce, and industry. Building such a system from so-called scratch is ludicrous because there are so many systems already in place that could be used to interconnect such a beginning system. This, of course, involves many so-called players in building such a system. Many of us have read in the papers about corporate mergers proposed to implement such a network in a given area. The role of the federal government as a guiding force promoting the formation of an information highway is the subject of much speculation and discussion by the telecommunications industry. A more detailed investigation of existing systems will be conducted in the ensuing chapters, and methods of interconnection will be explored.

2

Types of Wire and Cable Systems

Many people involved with the telecommunications community are very familiar with the wired versus wireless controversy. In many practical systems in use today, the two methods actually complement each other in many cases. Wireless systems such as cellular telephones, microwave radio links, and satellite-based networks often act as bridges interconnecting wired systems. A case in point is the telephone industry in the United States, which is an operator of many cellular telephone interconnects, long-haul microwave radio links, and satellite systems. Satellite systems are used to connect the various telephone systems throughout the world. Operators of many satellite systems lease out channel space for the various telecommunications company uses.

The maximum frequency bandwidth of wireless systems is continually expanding when newer transmission equipment is developed. Presently this bandwidth is about 50 GHz, which represents less than the theoretical bandwidth of one optical fiber. Therein lies the practicality of continued use of a cabled system.

2.1 Conductive Cables

2.1.1 Open wire lines

Open wire lines were the earliest type of wired communication system. An example of an open wire line is shown in Figure 2.1. Such systems were constructed of hard-drawn solid copper wire of sizes between #6–#10 AWG on pole plant with glass insulators. Wooden cross arms were used so more copper wire lines could be added as needed.

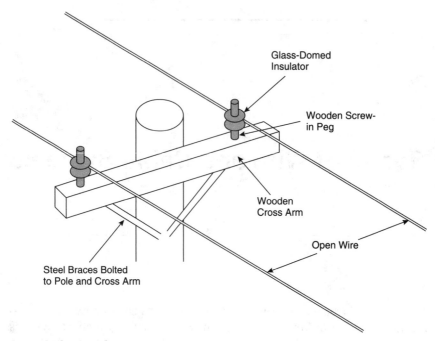

Glass-Domed
Insulator

Wooden Screw-
in Peg

Wooden
Cross Arm

Open Wire

Steel Braces Bolted
to Pole and Cross Arm

Figure 2.1 Open wire utility pole plant.

2.1.1.1 Telegraph (railroad) systems. Most of this early pole plant was placed along railroad rights-of-way and was originally intended to be used for railroad communications. However, the amount of time the system was used for railroad communications was small, and the excess time was used to carry commercial and domestic messages. The communication method was a manually operated telegraph that caused an electrical current to correspond to the dots and dashes of early Morse-type code. The speed of transmission and reception depended on the expertise of the operators. The Western Union Company was one of the largest and most successful telegraph companies and at one time enjoyed being the only long-distance message carrier.

Improvements in the keying (sending) and sounders (receiving) equipment enabled the speed of message traffic to increase. The teletype machine finally was developed, which virtually automated the telegraph industry.

Another user of open wire copper lines were the early fire alarm systems employing pull-boxes (transmitters) that signal the receiver at the fire station. This early system was electromechanical, where each pull box had a coded CAM corresponding to the location of the box. When the handle was pulled, a spring was wound, releasing the CAM, which in turn caused electrical contacts to interrupt the loop current in the line. The interruptions in the loop current was sensed by the receiver at the fire station and printed out on a paper tape to be read by the fire department personnel on duty. Often the loop current interruptions also would sound the fire department whistle or gong, calling fire department people to respond. Failure of the loop current signified a fault in the alarm system so repairs could promptly be made. Certainly this was a rudimentary system, but it was very reliable and many such systems are most likely still in use today.

2.1.1.2 Rural telephone systems. Since the technique of stringing open wire lines on utility pole plant was developed along railroad rights-of-way, it made sense to use the technique in wiring local communities for electric power distribution and the newly developed telephone industry. Since many communities had limited space along their road systems, room for only one pole plant was allowed. Common use of the pole plant by the electric power utility and the telephone company started the concept of joint pole plant ownership. The important working relationship between the telephone system and the power system was started nearly 100 years ago and has remained essentially intact ever since that time.

Like the fire alarm system, the early telephone system was quite rudimentary compared to what it is today. Each telephone set was connected to a bank of one or more manually operated switchboards (plugboards) located at a building housing what was known as the local exchange office. This building often looked like just another well-kept home in the neighborhood, complete with landscaped grounds. A call was initiated by a caller when lifting the earpiece of the telephone set off the hook, which caused the flow of a dc loop current to light a lamp on the operator's switchboard. The operator plugged a patch cord into the jack below the light and asked the caller what number was being called. Then the operator manually patched the lines together while causing the called telephone to ring. Once the caller or called party hung up their telephones, the light would extinguish and the operator could remove the patch cord connection. At this point in time, most telephone service was confined mainly to local calls for a flat monthly charge. Service from local exchange to local exchange was accomplished by connecting to a trunk line. Calls from exchange to exchange involving the trunk lines was considered as a toll call, which constituted an added charge on the caller's monthly bill. Indeed, this rudimentary service had wide acceptance, and the new telephone industry flourished.

High-quality, pleasant service with equipment and facilities designed for high reliability were the norm, and this caused the telephone industry considerable success. Since the telephone industry was a new enterprise, there were no people with experience to join the ranks. A good work ethic was about all that was to be had. Therefore, in-house training and/or on-the-job training was often provided by the telephone companies to build the installation and maintenance personnel crews. Also, the need for improved equipment caused the new telephone industry to promote company-sponsored research and development. The outgrowth of the research and development with new products and equipment caused the telephone industry to start its own equipment manufacturing. Thus the familiar Western Electric Company and the Bell System Laboratories started operations in the early 1900s.

2.1.1.3 Information capacity. The information-carrying capacity of open wire lines in the early days of the telephone industry was not a problem. The highest frequency currents were at about 3 kHz, the wire size was comparatively large, and the distances between the calling parties was not great. So it essentially was not a problem. As time went by, the pole plant was pretty much saturated with a maze of wires, which caused a maintenance problem for the telephone industry. Twisted-pair cable and a method of carrying several telephone calls on a single pair of wires was developed. This method required high-frequency carriers modulated with the telephone call information to be carried on a pair of wires. Now the telephone industry needed to consider the upper frequency limit of various classes and types of wires, which determined how many telephone calls could be carried on a single pair of wires. Even today in some remote parts of the country, there still might be some plant with open wire lines in operation. The upper frequency limit used on a good open wire system is nearly 150 kHz. According to telephone company standards, 12 voice channels can be served on an open wire system. The telephone industry philosophy was to research and learn as much about the system as possible. Continual testing, logging of data, study of outages, and maintenance records all contributed to the technical competency of the company.

2.1.2 Twisted-pair multiple cables

It became increasingly evident that improved telephone cables needed to be developed. Twisted-pair cables were developed back in the 1930s and 1940s. Twisted-pair cable allowed the telephone system to be expanded while not overburdening the utility pole plant.

2.1.2.1 Early cables. Early twisted-pair cable was made of solid soft-drawn AWG-26 wire, where each wire was insulated by paraffin-wax impregnated fabric, twisted with another wire, which in turn was twisted with other pairs. Color coding of the wires was done by colored threads contained in the fabric insulation. The whole cable was wrapped in paraffin-impregnated fabric and encased in a metallic lead jacket. This cable was hung from a galvanized steal messenger cable by wire loops spaced at close intervals due to the weight of the cable.

The cable was loosely hung, thus allowing the sag to adjust to the daily and seasonal temperature changes. This type of cable quickly replaced the open wire lines for local exchange service, which made for considerably better looking pole plant and allowed for expansion of telephone service. This type of cable was not without its problems. The constant movement caused by wind and temperature changes chaffed holes in the lead jacket. Rodents were known to chew and perforate the lead jacket. The holes allowed moisture and water to enter the cable and soak the fabric insulation, causing a low resistance path between the wire pairs. This low resistance connection between wires allowed a mixing of the telephone calls and created cross-talk, where a person would hear several conversations in the background. False off-hook conditions were another problem caused by excessive moisture. The repair problem was complicated and required the expertise of a person highly trained in working with metallic lead.

First the point of water entry had to be located before repairs could be made. The telephone company invented and manufactured several varieties of test instruments and procedures to locate the problem areas. Such equipment placed either a tone generator or pulser on one of the unused pairs in the cable. A person with an exploring coil on a long stick connected to a pair of headphones was used to walk the path of the defective cable. Since the metallic lead-covered cable would normally not allow any signal to escape. When the coil was getting close to the break, the repair person would detect a signal on the earphones. Once the break was located, the lead jacket was cut away, exposing the fabric-insulated wires. These wires were spread apart, and hot paraffin wax was poured over them, thus drying out the moisture. Then the wires were bundled together and wrapped in wax-impregnated fabric. The lead ends were cleaned and soldering paste applied. A larger split-lead cover was placed over the section and pounded down on the cable lead. Hot melted solder was applied to the cable ends and to the seam to complete the repair. This process often required several persons in the repair crew to perform the work and direct traffic. The temperature of the solder had to be correct. If not, a good seal between the lead jacket and the new cover would not be obtained, or the hot solder would melt away the lead jacket.

2.1.2.2 Present telephone cables. After the Second World War, many new plastics were developed. The new polyethylene, polypropylene, polyurethane, and polyvinyl plastics improved the quality of wire insulation to a remarkable point. The high insulating qualities allowed for thinner wire insulation, producing a more flexible cable. Telephone cable insulated and jacketed with the new improved plastic insulation solved many of the telephone industry's cable problems. The new cable did not absorb water and hence cross-talk all but disappeared. The cable presently used is much lighter and more flexible and thus allows the inclusion of more pairs of wires.

The tapping points are made in nonair or moisture-tight enclosures, allowing for simple connections to the wires.

Crimp-on splice connectors, often called *chicklets*, were developed, which increased the efficiency and speed of installation and repairs. This type of cable is available with its own built-in steel messenger wire and is often referred to as self-support cable, which is shown in Figure 2.2. Often for plant containing several telephone cables, a separate steel messenger cable is used to support the multiple cables, which are lashed together and to the strand with stainless-steal lashing wire. Modern-day twisted-pair telephone cables have improved the quality of the telephone plant to a high level of reliability and decreased maintenance costs to much lower levels. Most telephone company pole-line maintenance crews working on the lines are either making improvements or building new plant when not making repairs.

Aerial pole plant is quite common for electric-telephone distribution plant. However, the new plastic cable has been and is now manufactured for buried underground use. In many new housing developments and new industrial parks, the telephone plant is directly buried in the ground. Points of connections are made in vertical steel or plastic enclosures called *pedestals*. Customer telephone connections are made at these pedestals, and an example is shown in Figure 2.3. Where earth conditions are poor or in urban areas, underground plant is placed in conduits made of steel or plastic. The technique used is to blow with a compressed air source a shuttle with a thin plastic line fastened to it. A line can be blown through a conduit a distance of approximately ½ mile. Once this line is in place, it is used to draw a larger pull line through the conduit. In many instances several cables can be pulled through a conduit. Pulling lubricants are often used to ease the pulling operation.

2.1.2.3 Information capacity. Local area network (LAN) systems are used to connect computer systems to form an interoffice or interbusiness network. With the arrival of the personal computer, many businesses opted to

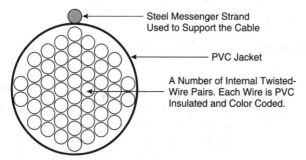

Figure 2.2 Self-supported multipair plastic insulated telephone cable.

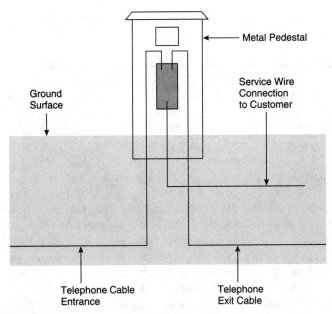

Figure 2.3 Example of underground cable enclosure.

plan equipment in the offices of certain workers to improve their work efficiency. It also became evident that information exchanged between offices
with computers (workstations) would be a big improvement in the work
flow. The simple solution was to hard wire certain workstations via a serial
or parallel input/output port. This solution was cumbersome, which paved
the way for some kind of communications control box—a modem. Telephone-type modular wire used to install ordinary telephones was selected
by several manufacturers of LAN systems. The modem is an important
piece of equipment and is often microprocessor-controlled itself. There are
many types of LANs with various designators. However, the transmission
medium is one of three types:

1. Twisted copper wire pairs.
2. Coaxial cable.
3. Fiber-optic cable.

Users of LAN systems should be very interested in connecting to an information highway because it will allow the benefits of computer-controlled communications with customers, vendors suppliers, and shippers. Many of the
computer manufacturers such as IBM and DEC have produced LAN equipment for their customers. To connect various LAN systems using different

methods, wires, and protocols will present a problem, which might indicate the need for interface equipment to a common communications system.

2.1.2.4 Local area network cables. The information-carrying capacity of so-called twisted-pair cable is indeed limited, and various users of twisted-pair cabling often use different criteria to specify maximum performance. The telephone system will rate the wire loop distance (wire path between called and calling parties) according to the limit of attenuation (where speech signals are weak) to loss of connection (depends on loop resistance). The transmission path should have sufficient loop current to keep the connecting relays closed (to keep the connection) and less attenuation so the talking parties can hear each other. For normal voice signals, the band of frequency the telephone system uses is 300 Hz to 3000 Hz. Loop loss is usually measured at a frequency of 1000 Hz in North America and is usually specified in decibels. Loop resistance of a wire pair is often specified in ohms per 1000 feet or ohms per kilometer or per mile. Wire sizes used in telephone systems and the resulting loop resistances are shown in Table 2.1. A discussion and derivation of the loop resistance equations is contained in appendix A.

The resistance is the round-trip resistance of a pair of wires. The loop resistance limit that allows the connection to be maintained for relay (electromechanical) controlled exchanges is about 1200 ohms. For exchanges using electronic switching, the loop resistance can be as high as 1800 ohms, and with the latest type of switches the loop resistance is extended to as much as 2400 ohms. If the maximum allowed loop resistance of an exchange is known, the preceding equation can be solved for the wire size for the connection route. If any section of the path lacks the proper wire size, then rerouting or placing a parallel conductor can solve the problem. Examples of using the formula are shown in Figure 2.4.

TABLE 2.1 Wire Size and Loop Resistance

AWG of wire	Attenuation in dB/km @ 1000 Hz	Loop resistance ohms/1000 ft	Loop resistance ohms/km	Loop resistance ohms/mi
19	0.75	16.1	46	85
22	1.00	32.5	92.5	171
24	1.30	52	148	274
26	1.65	83.5	237	435

The equation developed for the loop resistance in ohms per mile for various diameter wire is:

$$R \, loop = \frac{0.1095}{d^2}$$

R loop is in ohms
d is in inches

Example 1: For an exchange using an 1800-ohm maximum-allowed loop resistance, the loop resistance for #26-AWG copper wire is taken from Table 2.1 and is 435 ohms/mile.

$$\frac{1800 \text{ ohms}}{435 \text{ ohms/mile}} = 4.14 \text{ miles of plant}$$

Example 2: For a path of 5 miles, a larger-size wire will be needed. So:

$$360 = \frac{0.1095}{d^2} \quad \frac{1800 \text{ ohms}}{5 \text{ miles}} = 360 \text{ ohms/mile}$$

$$d^2 = \frac{0.1095}{360} = \frac{10.95 \times 10^{-2}}{3.60 \times 10^2} = 304 \times 10^{-4}$$

$$d = 3.04 \times 10^{-4} = 1.744 \times 10^{-4} = 0.01744 \text{ in.}$$

This is 17.44 mils and corresponds closest to #25-AWG cable. However, #24-AWG will be a better choice.

Figure 2.4 Example of loop resistance and distance calculations.

The line attenuation limit for a loop is 6 dB, and in the United States the limit might be as high as 9 dB. Of course, this attenuation is a function of frequency, and the preceding limits are considered the maximum. Usually this maximum loss is at the midband standard test frequency of 1000 Hz. Twisting the cable pairs tends to cause the inductive effects to cancel between pairs by essentially causing the mutual inductance to become negative. This is illustrated in Figure 2.5.

However, the close proximity of the two parallel conductors twisted together causes the capacity to increase substantially per unit of cable length. Adding inductances along the loop pair to counteract the effects of capacitance buildup is called "loading the line." Insertion of these series inductances, called loading coils, is spaced at fixed intervals to simulate a

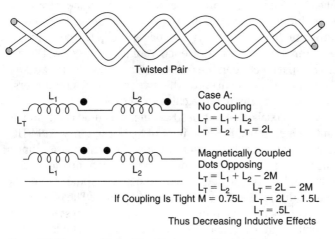

Figure 2.5 Decreased inductive effects caused by twisting.

so-called flat line. Loading with the series inductances does improve the voice frequencies carried, but since series inductive reactance increases linearly with frequency, this limits digital-signal use of the equalized line. Therefore, loading a line causes a decrease in propagation velocity and an increase in the line impedance. This relationship is proved in appendix B. Loading coils are added at various distances along the line so as to approximate a smooth transmission line. Therefore the total amount of needed inductance is more evenly distributed along the line.

The telephone industry uses a standard list of coded loading coils of various sizes of inductances and spacings between coils. Adding series inductances to the line improves the voice frequency transmission loss at the expense of higher frequency response. This higher frequency response loss prohibits higher frequency carrier transmissions and digital signal transmission, which limits the use of the cable plant. Twisted-pair type cables that are nonloaded have a practical upper frequency use of approximately 300 kHz. The telephone industry uses twisted-pair cables nonloaded, and using frequency division multiplexing (FDM) techniques is practically limited to 12 full-duplex voice-grade channels at distances up to 200 miles. Use of repeater amplifying equipment spaced along the cable route is necessary to overcome signal attenuation. Computer modems used to interface with the standard telephone system operate at data rates of 14,400 bps (bits per second), which corresponds to the CCITT (international) Standard V.33. Also, several types of LAN interbuilding wiring systems use telephone type twisted-pair cable. These short distances make it unnecessary to use any repeating amplifier equipment.

2.1.3 Coaxial cables

As the name implies, coaxial cable consists of a center metallic-wire conductor suspended within a metallic-tubing outer conductor having a common center. Polyethylene foam or polystyrene spacers keep the center wire on center with the outer sheath (tubing). Hence a concentric (coaxial) configuration results. The space between the outer and inner conductor acts as an insulator (dielectric), thus separating the two required conductors. Figure 2.6 shows the construction of a typical coaxial cable.

Prior to World War II, coaxial cable was used as shielded audio cable and in some early radio antenna systems. Development of high-frequency and VHF radio systems and radar systems also required improved transmission lines and cabling. Hence, coaxial cables were developed, which provided the means for connecting high-frequency equipment. The development of communications equipment during the war years was tremendous. The telephone industry was forced to expand the country's communication systems, and new long-haul telephone equipment and cable was developed. The Bell Labs were indeed working overtime. After the war, surplus equipment found its way to the civilian market. Early coaxial cable was of the braided, outer con-

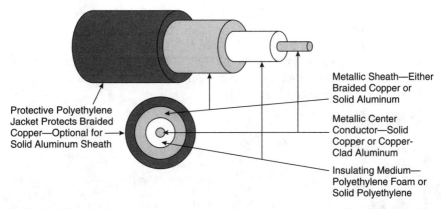

Metallic Sheath—Either
Braided Copper or
Solid Aluminum

Protective Polyethylene
Jacket Protects Braided
Copper—Optional for →
Solid Aluminum Sheath

Metallic Center
Conductor—Solid
Copper or Copper-
Clad Aluminum

Insulating Medium—
Polyethylene Foam or
Solid Polyethylene

Figure 2.6 Coaxial cable.

ductor type and appeared in sizes of ⅜ inch approximate diameter to ¼-inch diameter, quite similar to present-day RG8/AU, RG11AU, RG58, and RG59 types. It was this early, abundant source of quite inexpensive cable that gave the cable television industry its start. Characteristic cable impedance was either 50- or 75-ohms nominal impedance.

2.1.3.1 Cable television cable plants. The need for long-haul, efficient telephone service caused the telephone industry to research many methods. Twisted copper pair cable with its buildup of resistance and signal attenuation with increase of distance was indeed a difficult and expensive operation for the telephone industry. A means of carrying several duplex telephone channels on a single equivalent wire pair was developed. This method used a radio technique that employed high-frequency carriers modulated with voice signals. Placing several carriers on a wire, each carrying a one-way telephone channel, increased plant efficiency. Coaxial cable with its wideband width capability was the long-haul cable of choice. This cable plant was lashed to the ordinary utility pole plant containing local exchange service. However, it progressed to longer distances. Repeater amplifiers were spaced along the cable route to amplify the signal carriers attenuated by the coaxial cable. Research and development performed by the telephone industry resulted in developing the theory of cascaded amplifiers, which provided the information for the basis of cable television technology.

The significant electrical characteristics for any coaxial cable are its signal attenuation, which is a function of frequency, and its loop resistance. Both of these characteristics are a function of cable cross-section geometry and length. The approximate characteristic impedance value based on the cross-section geometry is illustrated in Figure 2.7. The attenuation of coaxial cable depends also on its cross-sectional size and is given in dB per unit of length. Larger size cable has lower loss than cable with a smaller cross section. The loss of any size cable increases with increasing frequency. The usual measure for cable loss is in dB per 100 feet or dB/100 meters and is often presented in chart or table form at various selected test frequencies. An example is given in Table 2.2 for standard sizes 0.5-, 0.75- and 1.0-inch diameter cable. An examination of the values given in the table clearly show that cable attenuation increases as the diameter decreases, and attenuation increases with increasing frequency.

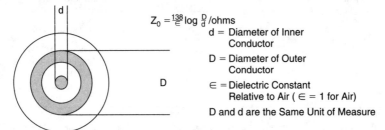

$$Z_0 = \frac{138}{\in} \log \frac{D}{d} / \text{ohms}$$

d = Diameter of Inner Conductor

D = Diameter of Outer Conductor

$\in$ = Dielectric Constant Relative to Air ($\in$ = 1 for Air)

D and d are the Same Unit of Measure

Figure 2.7 Approximate formula for characteristic impedance in ohms based on the cross-sectional geometry of coaxial cable.

TABLE 2.2 Coaxial Cable Attenuation for Three Common Cable Sizes

Freq. in MHz	0.500-inch loss @68° F in dB/100 feet	0.750-inch loss @68° F in dB/100 feet	1.0-inch loss @68° F in dB/100 feet
5	0.14	0.10	0.07
30	0.35	0.25	0.18
50	0.46	0.32	0.23
110	0.68	0.47	0.35
174	0.86	0.60	0.45
220	0.97	0.67	0.50
300	1.14	0.79	0.60
350	1.23	0.86	0.65
400	1.32	0.91	0.70
450	1.40	0.97	0.74
500	1.48	1.03	0.78
550	1.55	1.08	0.82
600	1.62	1.13	0.85
650	1.69	1.18	0.88
700	1.75	1.22	0.92
750	1.81	1.26	0.95
1000	2.09	1.46	1.10

An approximate formula for calculating cable attenuation between values given in the above table is as follows:

$$\frac{A_1}{A_2} = \sqrt{\frac{f_1}{f_2}}$$

A_1 = attenuation in dB at a lower frequency f_1

A_2 = attenuation in dB at f_2 at a higher frequency than f_1

Use of this approximate formula gives results within 1% of the true value.

High-frequency use of coaxial cables follows transmission line theory. Such theory states that the transmission line has to have the source or driving impedance and the load impedance accurately matched to its characteristic impedance. If these impedances are not matched, reflections occur along the line, which causes echoes or ghosts in the resulting signal. These reflections cause standing waves where the signal amplitude varies along points on the line. For bad mismatches, the signal will all but disappear at one-half wavelength intervals along the transmission line. The telephone industry developed the use of the concept of return loss, which is a measure of the logarithmic ratio of the incident signal power and the reflected signal power. This can be expressed either in terms of the impedance matches or the reflection coefficient. The common expression for return loss is given by the formula:

$$RL = 20\log \frac{1}{k}$$

k reflection coefficient

A derivation of this expression and a discussion of some key elements of transmission line theory are given in appendix C.

The loop resistance of coaxial cables is also of interest. In many cases, dc or low-frequency ac is placed on coaxial cable systems that are used to power repeating amplifiers along the transmission line. Since the center conductor is actually a lot smaller than the outer sheath, it is the principal cause of the loop resistance of a cable length. Often the outer conducting sheath is aluminum and the center conductor is either solid copper or copper-clad aluminum. The type of coaxial cable with the solid copper center conductor naturally has the lower loop resistance. Also, the larger size coaxial cable has a lower loop resistance. Table 2.3 shows the various values of loop resistance per thousand feet for the same type of cables in Table 2.2.

2.1.3.2 Cable television cable plants. The cable television industry was started after World War II when the television boom took the country. For many, reception of good television pictures was a problem. Television broadcasting transmitters were not very high in power, and an outside

TABLE 2.3 Loop Resistance in Ohms/1000 Feet at 68°F (20°C)

Cable-size diameter			Center conductor
0.50 inch	0.75 inch	1.00 inch	material
1.10	0.48	0.25	solid copper
1.50	0.65	0.35	copper-clad aluminum

rooftop receiving antenna was often necessary to receive decent television reception. For people living far away from a television station or in a valley, good television service was next to impossible. The cable television industry was started to provide service to such people living beyond the limits of television reception.

A receiving site selected either on a mountain or by the erection of a tall mast acted as a superreceiving antenna. Television channels were adjusted in amplitude, combined, and fed to subscribers via a cable-connected cascade of amplifiers. Early amplifiers were rudimentary in design and often home-built. Cable was mainly war-surplus coaxial cable strung on the utility poles. Problems with permits, rights-of-way laws, and utility pole attachment agreements were some of the nontechnical problems facing the fledgling industry.

In the beginning, cable television operators provided service on standard broadcast television channels so subscriber television sets could tune in the selected station. For some cable systems, a selected VHF television station on, for example, channel 13 (216 MHz) would be attenuated by the cable a lot more than channel 2 (60 MHz). Therefore, at the receiving site, channel 13 would be received on channel 13 and then converted to a lower, unused off-air channel, channel 3 (66 MHz), for instance, to be transmitted along the cable system to subscribers on channel 3. Many early systems were only 5-channel systems with service available on only channels 2–6. Tapping into such early cable systems for subscriber service was a difficult task, depending on the type of cable.

Transmission line theory says that the tap system should not upset the cable impedance match, which could cause standing waves on the line and hence voltage-level variations at points along the line. If the mismatch is great enough, the signal at various frequencies all but disappear at various points along the length of the line. If at one of these points a subscriber is connected, the signal on some channels could be far too small for acceptable television reception. Such problems, along with the public's acceptance of the cable television business, caused the cable operators and the equipment manufacturers to produce improved products and services. One of the best improvements was better cable and connectors. The seamless aluminum sheath cable was constructed with plastic foam insulation and a copper clad center conductor. In many areas such as away from the seashore with a salt air environment or industrial area with corrosive air pollution, a covering plastic protective jacket was not needed. System design concepts were also improved, and the newly developed cable had acceptable loss limits permitting operation to channel 13, allowing a full 12-VHF-channel cable system.

Cascade amplifier theory developed by the telephone company and experience gained by the cable engineers and technicians produced many design improvements for the cable television industry. Instead of tapping into a cable at a new subscriber location, the idea of a dedicated tap system was conceived. Such a system was built with a tap port dedicated for each prospective subscriber home or lot of land. The tap value and location was initially built into the design so the signal level at the various channels could be predicted. The concept of trunk and feeder was developed where the trunk cable was a separate transportation system containing no subscriber taps. Essentially, the trunk branched from the central receiving site to the various ends of the system using a tree-and-branch network architecture. At trunk amplifier locations, a bridging amplifier circuit would isolate the trunk system and provide signal to the shorter subscriber feeder amplifier cascade containing the subscriber tap devices. The 12-channel systems employing this architecture provided service to subscribers on the 12 standard television channels that could be received by the subscriber's television set.

Enter UHF service, which operates in the frequency band of 470–890 MHz. The newer television sets employed a separate dial or channel selector for UHF and a separate set of antenna terminals for a UHF receiving antenna. What was intended was for cable subscribers to install a rooftop or separate UHF antenna to receive UHF stations and receive them off-air and not from cable. UHF broadcasting transmitters had to be of much higher power than VHF broadcast equipment, and even then the over-the-air attenuation was often large enough to cause poor reception in many areas. Also, the attenuation of the coaxial cable was so large that it prohibited carrying UHF service directly on the cable. Cable operators appeared stuck between a rock and the proverbial hard place.

A study of the spectrum of a 12-channel VHF cable system indicated a vacant spot between VHF channel 6 and VHF channel 7 (88 MHz–174 MHz). Over-the-air signals within this frequency band were occupied by the FM broadcast radio band, the aircraft communications band, commercial communications band, and an amateur radio band, all of no interest to cable subscribers. In many cases the FM radio band was carried on the cable as a separate service. Now subscribers could receive a good quality signal and not need any antenna equipment. Cable operators found that 9 standard 6-MHz television channels could fit in this space above the FM band at 150–174 MHz. The only problem was that a standard television channel selector could not tune in any of these channels.

Hence the first subscriber add-on device known as a converter was developed. Cable equipment manufacturers were in general very close to the needs of the industry and developed a converter that would receive these signals and translate them to appropriate vacant UHF television channels. These cable channels appearing in the so-called midband between the FM band and channel 7 were labeled originally as cable channels A–I. The converter box would translate them to standard UHF channels where the subscriber could tune the desired channel in on the UHF selector dial. This converter box was often called a block converter because a block of midband channels was converted in a block to the UHF dial. The method employing a block converter is shown in Figure 2.8. Cable television systems

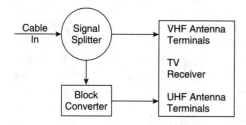

Figure 2.8 A block converter filters out midband signals and translates block of midband signals to standard VHF dial channels.

using this technique were able to offer their subscribers a maximum of 21 television channels, plus FM radio service from a cable plant originally carrying the 12 ordinary VHF standard television broadcast stations. Such cable systems were referred to as 21-channel systems.

Expansion of cable television systems above channel 13 (220 MHz) became more and more appealing as better cable and improved repeater amplifiers became available. Since cable loss or attenuation increases with increased frequency, amplifier spacing became less, i.e., closer together. Thus more amplifiers were used per mile of plant for operation at higher frequencies. The next milestone achieved by cable television technology was the 300-MHz 36-channel systems followed by 400-MHz 52-channel systems. Today many systems have an upper frequency band of 750 MHz with approximately 110 TV channels. Some systems are planning for an upper frequency goal of 1000 MHz, which would allow carrying the UHF band unchanged. Amplifier spacing for such a system would be quite small, which means many amplifiers per mile. This in turn would use a lot of power and increase maintenance. Practically, high-frequency cable systems would be quite limited in geographical size.

The amplifier cascades in cable systems are limited in number, i.e., how many can be used in a cascade. Noise and distortion buildup caused by the cascade of amplifiers is limited so as to limit any noticeable picture impairments. Amplifier cascade theory developed by the telephone industry was used and developed by the cable television industry. It was found that distortion buildup, either second- or third-order distortion, depended on the number of television channels the amplifier cascade had to amplify. It was also discovered that with slight adjustments in frequency, the distortion elements would still build up but would not cause visual picture impairment. This change in frequency, although slight, still caused tuning problems for some television receivers. However, since such systems employed tunable converters, television tuning was not a problem. It should be noted that 36, 52, etc., cable systems had to employ converters to translate the whole cable service to a single channel tuned by the television set. Such a converter arrangement is shown in Figure 2.9.

Interior cable wiring in a subscriber's home used small drop-coaxial cable of the RG59 or RG6 75-ohm variety. The upper usable frequency limits of this small size cable is about 600 MHz due to the excessive signal attenuation. The signal bandwidth of cable television systems is much greater than is possible using twisted-pair telephone-type cable. However, such cable systems were essentially a downstream, one-way system to usually residential areas where subscribers lived. Delivery of good television service was the stock-in-trade for cable television systems.

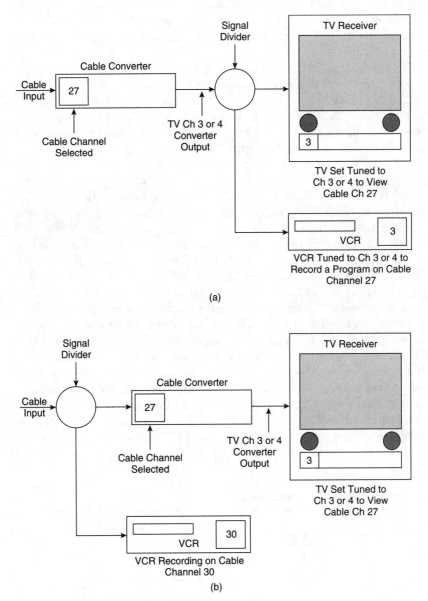

Figure 2.9 Sample cable-to-TV and VCR interconnections.

In some metropolitan areas, a requirement for upstream or two-way (duplex) service was specified as part of the licensing or franchising agreements between the municipality and the cable operator. Since the lowest television channel operates at 55 MHz (channel 2), the cable spectrum below this frequency was vacant, thus allowing a few television channels to be placed in this space and head upstream. The band of frequencies was 5 to 30 MHz, thus allowing 4 television channels of 6-MHz per channel to be carried in this space. The band between 30 and 55 MHz was unusable due to the crossover (diplex) filter separating the forward and reverse amplifier cascade.

This system was referred to as a *subsplit reverse system*. Since the cable loss at these low frequencies was small, fewer amplifiers were needed in the reverse cascade than in the forward cascade. The reverse amplifiers were usually installed with their diplex filters in the same housings as the forward amplifiers and if a reverse amplifier module was not required, a simple jumper was used at that amplifier location.

It should be evident that there were many more downstream channels than upstream channels for a subsplit two-way system. When more upstream service was required, the cable television industry developed the midsplit and high-split systems. These systems had many more reverse channels but were never at a 1:1 ratio.

Table 2.4 describes the forward/reverse frequency bands for a system with an upper frequency limit of 450 MHz. Two-way systems, if they were built at all, appeared in dense metropolitan areas. Many municipalities that required such systems to be built soon discovered that to use them, additional personnel, equipment, and added expense were needed. Some municipalities required a separate bidirectional cable system to be constructed for use by the municipality. These systems were either a single cable with a subsplit, midsplit, or high-split bidirectional configuration or a separate cable amplifier cascade for each direction. Such systems were often termed institutional networks or "I" networks and did not carry standard commercial television programming. If these systems were used at all, they carried surveillance camera video, alarm data, school interconnect television, and/or some computer data. Unfortunately, such systems were constructed and maintained by the licensed cable television company at considerable expense.

TABLE 2.4 Two-Way Cable TV Systems

	Subsplit		Midsplit		High-split	
	no ch	freq MHz	no ch	freq MHz	no ch	freq band
forward	60	55–450	50	150–450	36	234–450
reverse	4½	5–30	18	5–112	28	5–174

2.1.3.3 Local/wide area network systems.
Local area or wide area data networks (LAN/WAN) were developed in conjunction with the widespread use and acceptance of personal and business computers. As discussed earlier, many such systems used twisted-pair telephone-type cable. However, as the necessity for high-speed data became apparent, improved cable systems were developed. Many types of LAN/WAN networks' control equipment employing software and hardware were developed often by the manufacturers of computers and data equipment. Digital Equipment Corporation (DEC), Intel Corporation, and the Xerox Corporation jointly worked on a system that became known as the Ethernet specification and used coaxial cable similar to video and cable-television drop cable as the transmission medium.

Such cable will handle computer communications at speeds of 20 megabits per second (Mbps) over short distances directly. This conforms to IEEE standard 802.3. This standard describes the communications protocol between communicating computer workstations. Most Ethernet systems use a specified cable type and entry port, often a wall plate with a connecting jumper to the communications modem. Coaxial cable is run through building walls and conduits to provide a communications highway between the corporate offices and plant.

In many instances, commercial power system noise and the electrical noise caused by manufacturing machinery can be a major cause of data errors in computer equipment. Coaxial cable with its outer conductive sheath acts as a shield against such noise. However, if the noise sources are great enough, noise ingress will still result and different methods will have to be used. One such system is termed MAP/TOP (Manufacturing-Automation-Protocol/Technical Office Protocols). This is a protocol of seven layers that set the system controls for transfer of data through a transmission medium. Although the transmission medium is essentially not specified by the standard, a broadband coaxial cable cascade of amplifiers, one for each direction, have been used very effectively in large manufacturing plant environments.

To implement a MAP/TOP system, operators chose a broadband cable television technique using a single cable of either a midsplit or high-split method. An in-plant, large bandwidth system could support video/audio-training-developed information, in-plant video security surveillance, and telephone intercom, paging, and digital data interchange using the MAP/TOP protocols. Systems constructed in this fashion operated very well in electrically noisy manufacturing climates. General Motors Corporation used this system in some of its automobile manufacturing plants and endorsed its use over some of the other coaxial cable methods such as Ethernet techniques as promoted by Digital Equipment Corporation and others.

In theory, the MAP/TOP backbone coaxial cable system could support significantly higher data rates such as the telephone industry's T-1 standard, which has a bit rate of 1.544 Mbps. One benefit of using a single cable bidirectional cascade of amplifiers was to take advantage of the large amount of highly developed cable television equipment available on the market. The installation of such systems used the same connectors, taps, couplers, and amplifiers as did the cable television systems. This resulted in significant cost savings to the user. Constructing this type of system for industrial purposes resulted in significant added business to cable television equipment manufacturers and contractors.

2.2 Nonconductive cables

So far, all the cables and wires used in the telecommunications industry are essentially conductors of electricity, i.e., made of conductive metals. If one thinks about a direct current (varies at zero frequency) that flows along a conductor to an end or load point, the concept of electric transmission is realized. If this electric current is varied at a period, i.e., rate, it now varies so many times a second that the same thing happens. Electricity is transmitted through the cable. Now if the frequency is increased, the electric energy transmitted tends to flow on the outside of the conducting wires, hence some is lost in the space surrounding the conductors.

Coaxial cable is a much superior method used to transmit very high frequency electric current. This is because the outer conductor is essentially tubing, and the electric and magnetic field is confined inside the cable. As earlier stated, this loss increases with frequency, where the loss at extremely high frequencies, that is greater than 1 gigahertz (1000 MHz), makes use of coaxial cable impractical. Radar technology employing frequencies well into the gigahertz range employ a waveguide principle that guides the electric and magnetic fields toward the load end. Waveguides are essentially like pipes with carefully designed dimensions that only allow a very narrow band of these high frequencies. The study of physics has proved that what we know as light is actually extremely high frequency electromagnetic energy (electricity) where the wave travels through space. Also, white light contains all the colors (frequencies) in the visible spectrum. The human eye is a detector of light in the visible spectrum, and as white light is reflected from objects, this causes certain frequencies to vary in amplitude. The eye detects this image in the form of shapes and colors. The frequency of this light energy is between 10^{14} Hz to 10^{15} Hz. Frequency is related to wavelength and the speed of electromagnetic wave propagation in the following mathematical expression:

$$\text{Wavelength}/_m = \frac{\text{speed of propagation } (3 \times 10^8 \text{ m/s})}{\text{frequency in Hz}}$$

Figure 2.10 shows the visible spectrum of colors as a function of the wavelength. If solved for frequency, above relationship shows that as wavelength increases, frequency decreases. Hence the color red at a wavelength

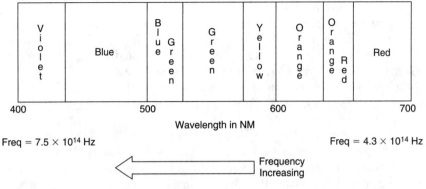

Figure 2.10 The visible spectrum.

of 700 nm is a lower frequency than violet at a wavelength of 400 nm (nanometer, a measure of wavelength in the optical region).

The use of the electromagnetic spectrum at light-wave frequencies for communications purposes was researched and investigated by many. However, the telephone industry, namely Bell Labs, was the most notable.

2.2.1 Light pipes/conduits

Early light-wave communications systems used what was known as light pipes. Rods of transparent material such as glass, Lucite, or other plastic material was bent into shapes and lengths to transmit light through the pipe. Piping light to difficult, small, and confined areas found use in industrial inspection and in some medical applications. The automobile industry used light pipes as turn-signal indicators that appeared, and still do, in several models of automobiles. The light-pipe ends appeared at the top of the front fenders visible from the driver's seat. When observed during operation, the driver actually saw a sample of the light emitted by the turn signal lamps at the display on the fender. In most cases, light pipes were covered by a plastic jacket to protect them from breakage and scratching. Often, light pipes had to be made in sections that were connected mechanically or cemented with transparent cement. If the light pipe had a smaller diameter or smaller cross section, more flexibility was obtained and hence installation and maintenance was simplified. Light pipes essentially were operated with light in the visible spectrum.

2.2.1.1 Optical fiber cable. Optical fiber was a natural outgrowth of light pipes. Studies of transparent material indicates that at the visible spectrum, some plastics have less light attenuation than glass. Long, round, thin strands of glass are actually quite flexible and can be drawn out to tremendous lengths. Studies of the loss of light attenuation of various types of glass as a function of frequency have produced some interesting results. The Corning Glass company did a lot of research on the loss of silica (glass) fiber during the 1970s. Measurement of loss is measured in dB of light power per kilometer and is often referred to as the loss coefficient.

One of the earlier problems was how to maximize the amount of light into a fiber. One technique was to clad or cover the fiber with a glass material that had a slightly lower refractive index. Also, light entering the end of a fiber had to be within the cone of acceptance for the particular glass fiber. An example is shown in Figure 2.11. The cladding guides the light rays falling within the cone of acceptance down the core and, being reflected, into the core by the cladding. If the cladding material has a constant index of refraction through its wall thickness, the fiber-optic cable is known as *step index fiber*. If the cladding material has a gradual change in refractive index, this type of cable is known as *graded index cable*. Both types are also known as multimode cable. The light at various frequencies propagates down the cable at various modes.

Multimode cable has a large-diameter fiber core. Research on fiber-optic cable has shown that if the core size is as small as the wavelength of the light, the wave propagates down the fiber in one mode. Such cable is referred to as *single-mode fiber-optic cable*. For most communications applications, single-mode cable provides the lowest attenuation. Because the single-mode fiber has such a small diameter, it has a small aperture. Figure 2.12 shows a curve of attenuation versus wavelength of light for a typical silica glass single-mode optical fiber. The lowest attenuation falls in between 1300 and 1600 nm of wavelengths and is referred to as the window area of

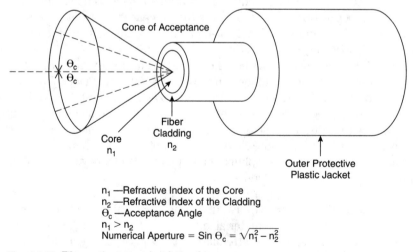

n_1 —Refractive Index of the Core
n_2 —Refractive Index of the Cladding
Θ_c —Acceptance Angle
$n_1 > n_2$
Numerical Aperture $= \mathrm{Sin}\ \Theta_c = \sqrt{n_1^2 - n_2^2}$

Figure 2.11 Fiber-optic geometric relationship.

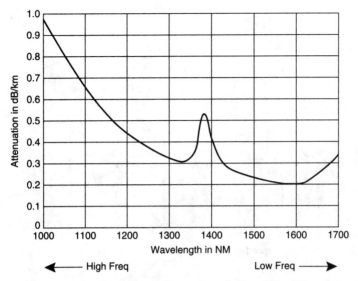

Figure 2.12 Graph of loss (attenuation) versus wavelength for a glass single-mode fiber.

low optical loss. Recall that 20 dB of coaxial cable at 1-GHz (1000 MHz) for 1-inch coaxial cable calculates out to a distance of 54 meters or 0.054 km. For the optical fiber, 20 dB of loss calculates out at its lowest loss point of 0.2 dB/km at 1550 nm to a distance of 100 km. In round numbers, the optical fiber has about 200 times better loss than coaxial cable. Of course, bandwidth is another matter.

To make a fiber-optic transmission system practical, a substantial light generator is needed to transmit a strong light signal down the fiber cable. Earlier multimode fiber-optic systems used a modulated LED transmitter and photodiode receivers. At lower data rates and shorter distances, this type of system worked well. The light from LED transmitters would travel along the optical waveguide in several modes. Light rays traveling along multimode fiber follow different paths and so arrive at the end at slightly different times. If sharp pulses control or modulate the light source intensity at the sending end of a fiber, the received light as detected by a photo diode or photo transistor will result in a much broader pulse due to the different path lengths. The type and geometry of the cladding has an effect on the broadening of the received pulse.

Figure 2.13 illustrates the effect. A gradual increase in refractive index of the cladding to the core lessens the effect. This type of fiber is called *graded-index fiber.* As shown in Figure 2.13, the graded-index fiber has less broadening of the receive pulse. Since digital signals are often transmitted as a series of pulses where a high level corresponds to a binary one and a zero level corresponds to a binary zero, any difficulty in being able to prop-

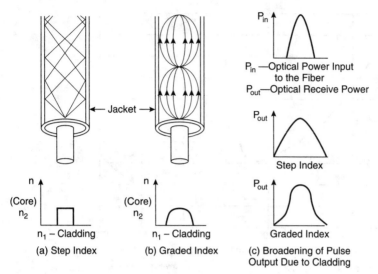

P_{in} —Optical Power Input to the Fiber
P_{out}—Optical Receive Power

Step Index

Graded Index

n (Core) n_2

n_1 – Cladding

(a) Step Index

n (Core) n_2

n_1 – Cladding

(b) Graded Index

(c) Broadening of Pulse Output Due to Cladding

Figure 2.13 Effects of fiber-optic cable cladding.

erly detect a one or zero level at the receiver will cause errors. Such errors are referred to as the bit error rate, BER, which corresponds to the number of errors per unit of time. Mathematically, the dispersion D is given by:

$$D = \sqrt{\frac{t_2^2 - t_1^2}{L}}$$

L is the length of the fiber (km)
t_2 is the output pulse width (ns)
t_2 is the input pulse width (ns)

The pulse performance of the graded-index fiber is significantly better than for the step-index fiber. There are three characteristics of optical fiber that cause dispersion:

- Material dispersion
- Modal dispersion
- Waveguide dispersion

The material dispersion depends on the composition of core material and is frequency dependent (color). The modal dispersion is controlled by the dimensions of the core. Waveguide dispersion varies with color (frequency). Multimedia fibers have small waveguide dispersion, whereas for single-mode fiber it might become significant. On the other hand, multimedia fibers have significant modal dispersion, while single-mode fiber does not. Single-mode and multimode fibers both have significant dispersion due to the type of material.

Installation of fiber-optic cable should be performed according to the manufacturer's specifications. Severe bending can cause light energy to escape out of the core, resulting in higher-than-normal losses. Most manufacturers of fiber-optic cable test each fiber in the cable for optical loss. If the manufacturing process becomes faulty, the fiber core in the buffer tube can have micro bends. Microbending can cause a fiber to exhibit high loss. Figure 2.14 shows how microbending takes place, which is the snaking caused by compressing the fiber in the outer jacket. Both single and multimode fiber cable can suffer from either bowing (macrobending) or internal bending (microbending).

The type of fiber-optic cable that is most used for communication purposes is single-mode fiber. The modal dispersion is negligible, and the material and waveguide dispersion is minimized by the choice of the glass material and the proper manufacturing process. Single-mode fiber has a very small cross section and a thin cladding of lower refractive index material. Losses are on the order of 0.3 to 0.4 dB/km. Cable can be obtained on reels of several kilometers. Splices of fiber-optic glass strands are usually performed with a fusion splicer. This type of splicing, of course, has to take place in the field, and since the process is very exacting and cleanliness a necessary priority, an elaborate trailer or truck vehicle laboratory is used.

A good fusion splice results in a typical splice loss in the range of 0.02–0.1 dB. Manufactured optical connectors are available from several manufacturers in a variety of types. Most connectors have optical losses of about 0.05 to 0.08 dB. Connectors have a short pigtail of about 5 to 10 meters in length. Incoming fiber-optic cable to a terminal point office or communication switching exchange usually is terminated in a splice tray mounted in equipment racks. Here the pigtail ends with connectors are fusion spliced to the fiber-optic cable strand.

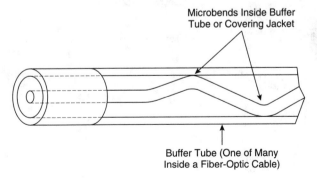

Figure 2.14 Fiber microbending.

Several types of recently improved mechanical splices have appeared on the market with losses approaching fusion splices. Most mechanical splices employ mating connectors for each end of the fibers to be spliced, and they employ quite elaborate techniques in preparing the fiber ends for the splice. Some types of mechanical splices use cement or gel material at the splice point. For a low-loss splice, proper fiber cleaving and cleaning is required. In general, splicing and connecting fiber-optic systems is a more difficult and expensive process than electrical cable systems. However, the extremely wide bandwidth and low loss are well worth the effort.

Fiber-optic cable is commercially available with a number of actual optical fibers in a single cable. Often several fibers with color-coded jackets covering the cladding will be placed in color-coded buffer tubes. A cross section of a typical fiber-optic cable is shown in Figure 2.15. Some varieties of fiber-optic cable have a metallic wire that is often placed at ground potential and can be used with metallic cable tracers to locate the buried cable. Fiber-optic cable is available from a few fibers up to 2000 fibers, depending on the application. Each fiber has tremendous signal bandwidth. Simple arithmetic makes clear the enormous signal-carrying capacity of large multifiber-optic cables. This will become increasingly clear in the later chapters of this book. A quick computation from Figure 2.12 says that a wavelength of 1700 to 1000 nm gives a frequency span or bandwidth of approximately 124 THz.

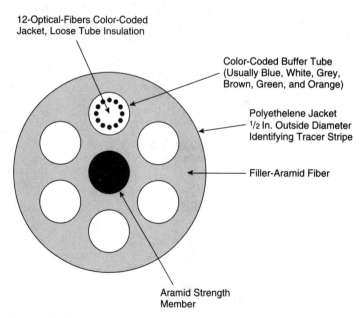

12-Optical-Fibers Color-Coded
Jacket, Loose Tube Insulation

Color-Coded Buffer Tube
(Usually Blue, White, Grey,
Brown, Green, and Orange)

Polyethelene Jacket
1/2 In. Outside Diameter
Identifying Tracer Stripe

Filler-Aramid Fiber

Aramid Strength
Member

Figure 2.15   Example of a fiber-optic cable cross section.

2.3   Routes, Conduits, and Pole Plants

The present communications lines that can contribute to an information highway follow many paths, highways, and byways. Most of us know and see everyday the utility pole lines, microwave radio towers, and satellite communication antennas. Physical cables and wires are either strung on pole lines or buried along the public streets and rights-of-way.

2.3.1   Utility pole plant

Most of us are well aware of the utility company pole plant commonly referred to as telephone poles. These pole lines appear along the local streets and roads connecting communities to communities. These poles carry the electric and telephone wires and cables serving customers along the way. The electric plant is connected to a local switching station, which in turn connects each community or municipal area to the power distribution grid. The telephone plant consists of local customer lines and trunk lines connecting the local exchanges to the higher class switching locations. In municipal areas, the telephone plant can become large, with many pairs of cables lashed to heavy steel supporting cables. In recent years, the telephone industry has been installing lighter weight and highly efficient fiber-optic cables. A single strand of fiber-optic cable can do the work of many twisted copper wires. Such cables operate as trunk lines and/or long-haul service lines. It is this fiber-optic cable plant that can contribute to the information highway.

2.3.1.1   Telephone/electric utility plant. By common agreement between the pole owners (the electric power company and the telephone company), the electric power transmission and service distribution wires are installed at the top of the pole. Often the very top wire is at ground potential and is referred to as the static line. This protects the current-carrying conductors of the primary and secondary lines from lightning. The high-voltage primary line is placed further down the pole length an appropriate distance (as determined by the voltage level according to rules of the National Electrical Safety Code). The primary lines might consist of one single-phase wire or three three-phase lines. Further down the pole, the electric distribution wires are placed, which again might consist of a single hot wire and a ground wire or of three three-phase wires and a ground wire. The spacing between this group of wires and the primary wires are according to the National Electric Safety Code. However, there are a multitude of various arrangements of electric power lines on utility poles that always occupy the top section of the pole. Normal domestic service for most homes is 220/110-Vac single-phase service, which consists of two hot wires and one ground wire leading from the utility pole to the customer's home.

2.3.1.2 Municipal and tenant use. Usually, 41 inches from the electrical distribution service wire is the space reserved for use by the municipality. The municipal fire alarm cable system is what usually occupies this space on the utility pole. Also, cable television service follows approximately 1 foot below this area and 1 foot above the telephone cable plant, which occupies the bottom area of the pole. Proper clearance above the ground is specified by the pole agreement rules and usually refers to the telephone system. The height of the pole has to be large enough so that all the services can maintain the proper spacing between each other. The telephone industry, which is usually noted for its in-house training, has essentially redone the written rules of the National Electric Codes into an easy-to-understand work pertaining to overhead and buried plant. This book is known in the industry as the Blue Book. In many cases this book is available to the power companies (which are joint owners of the utility pole plant) and the cable television companies (which are tenants on the pole plant).

2.3.1.3 Cable television tenant use. When a third-party tenant such as a cable television operator makes application with the pole owners to install its cable plant on the utility pole line, an actual three-party survey is made. This survey is an on-site inspection of each pole to ascertain whether adequate space for the cable operator's plant is available. A measuring pole is used to actually measure the spacing of the electric, telephone, and/or fire alarm lines to see if proper space is available. If, for example, all present occupants of the pole are within specification and there is no available room for the cable television applicant, then a taller pole is needed. The cable operator usually is required to pay for the new pole and the transfer cost of the lines and wires. In many cases, poles are tall enough for the addition of cable television plant. However, if the pole is tall enough but the existing telephone and electric plant are spread too far apart, then these existing wires will have to be rearranged to make proper space for the additional cable television plant. Again, the cable television operator has to bear the cost of this work. Also, the cable television operator has to pay each utility company for the cost of the survey and engineering costs associated with the changes agreed upon.

Cable television plant consists of coaxial cables lashed to a steel messenger-supporting strand in a fashion similar to the telephone plant. Both the telephone plant and cable television plant might use what is known as self-support or figure-eight cable. Essentially, this type of cable has its own integral steel messenger strand. Usually, smaller diameter cables are of this type. Cable television systems use what known as a trunk/distribution technique, which consists of a trunk cable and a separate distribution cable. The distribution cable has the subscriber taps installed at intervals near subscribers' homes, where drop wires are used to connect subscribers to the cable service. The trunk lines connect the cable service to the distribution cables via bridging/isolation amplifiers. Low-voltage alternating current is carried on the trunk cables and some distribution cables to power the line amplifiers. Most cable plant consists of at least one distribution cable to several distribution and trunk cables.

Many cable television operators like the telephone companies have been installing fiber-optic cables, which act as long-distance trunk cables. The low signal loss characteristics make fiber-optic cable extremely attractive to cable operators. In many cable television applications, a single fiber can carry the whole cable television service band the same distance as a cascade of many cable amplifiers with only a single optical transmitter/receiver pair.

2.3.2 High-voltage transmission lines

The high-voltage utility lines are the power lines connecting the national power grid. Huge, tall steel towers or metal poles are seen all over the continental United States, Canada, and Mexico carrying an array of cables connected by insulators. Some of the voltage levels are extremely high, ranging in typical values of several hundred kilovolts. Most generating plants generate initially high voltages on the order of 25,000 to 40,000 volts, and banks of step-up transformers increase the voltage level to high-voltage transportation lines. These lines are connected via the switching stations to the power grid. In many cases, the switching system is in a remote area and is essentially controlled remotely. Each electric company has a control station that monitors the flow of electric energy from the power grid to its own distribution system. Some power companies do not own any generating plants, and they buy their power from the grid. Other companies have extensive coal, oil, gas, nuclear, or hydroelectric generating plants and a large distribution plant.

2.3.2.1 High-voltage line signaling.
Signaling along the high-voltage lines was used at one time and essentially consisted of monitoring and/or line faults. As voltages along these lines increased, the difficulty of online signaling increased. Also, as the power grid system was expanded and more switching stations were needed, the control signaling needs increased as well.

2.3.2.2 Rights-of-way paths.
The paths followed by the high-voltage power distribution grid is through rights-of-way agreements that the many power companies have arranged throughout the years. These rights-of-way allow for plant expansion, tree and brush cutting, or a chemical defoliation program to keep clear the lines of overgrowth. Access by truck and inspection vehicles are necessary to inspect and monitor the lines. These rights-of-way paths are as long and straight as can be arranged and do not usually follow roads, except in many cases along major highways or railroad rights-of-way.

2.3.2.3 Communications and line control. As mentioned before, operation of many switching stations are by remote control, which requires some communication systems to connect the switching station to the controller. Microwave radio systems have been used to provide this control link and high-frequency packet radio systems. Most of the controlling signals are digital and are radioed in a high-speed packet back to the controlling station. When interrogated by the controller, operational data is obtained from the switching station as to power flow and switching-position information. Such systems are adequate at present. However, the power companies are looking at future communication needs and are installing fiber-optic cable along the power distribution rights-of-way. The information-carrying capacity of such fiber-optic cable is very much larger than the information needs of the electric companies. Hence the extra fiber capacity can be leased to any communication company. Now some of the electric power companies also can become players in the information highway.

2.3.3 Pipelines—oil and gas utility

The electric power companies are not the only companies that have long-distance rights-of-way. For many years the oil and gas industry have built a web of pipelines through the country. As the oil and gas needs of the country increased, larger pipelines were added. It is these rights-of-way that are of interest to the telecommunications industry.

2.3.3.1 Communication conduits. Along sections of some pipeline rights-of-way, some communication facilities were installed as joint agreements or ventures. Such communications facilities consisted of microwave radio links and some pole lines carrying telephone and commercial information services. Since pipelines were buried, some communication conduits were installed and buried alongside the oil or gas pipeline system. Various types of communication cables such as twisted-pair multiconductor telephone-type cable and/or coaxial cable and fiber-optic cable were installed in such conduits. Naturally, the various signals needed to control the oil or gas system were carried by some of these cables. The direct routes often followed by these utility pipelines resulted in shorter communication distances and the associated lower signal attenuation.

2.3.3.2 Uses for abandoned pipes and conduits. As mentioned before, larger pipes were often later added as the needs for energy increased. Sometimes this added plant was rerouted if it proved beneficial. Unused and/or abandoned pipelines that became too small for the pipeline operator were left in place. One company that was a pipeline operator supplying oil and gas from the fields to users in various parts of the country decided to enter the communication business. By selecting sections of abandoned pipes, a path was connected cross-country, and fiber-optic cable with many fibers was installed. Communication engineers and technicians were hired to test out the fibers and assist customers in making connections. As an off-shoot of the Williams Company, an oil and gas pipeline company based in Tulsa, Oklahoma, the Wiltel company became a long-distance communications network operator. This fiber-optic system supplies telephone, video, and data communications services to the many Wiltel customers, and this type of innovative approach to communications is the very tonic needed to produce a truly cost-effective information highway.

2.3.4 Railroad lines and paths

Probably the oldest rights-of-way paths in the country were the railroads. Since rail transportation was integral in the development of the country, the United States government aided wherever possible the development of the railroads. Obtaining the rights-of-way was indeed very necessary in extending the country's railroad system. Eventually the railroads were extended east-west and north-south, connecting cities and industrial areas. When the telegraph was developed, the railroad system had its own telegraph lines so train arrival and departure information could be sent from station to station. Also, the railroads became a major carrier of the U.S. mail service. Pole lines with telegraph wires were strung along those railroad rights-of-way, and the telegraph industry (mainly the Western Union Company) flourished.

2.3.4.1 Leased rights-of-way. It was these rights-of-way along the side of the railroad tracks that many railroads leased out to common communication carriers. Western Union telegraph and telephone companies either leased a right-of-way or space on the existing pole lines to carry their communication facilities. This, of course, provided added revenue for the railroad operators. Also, some telephone companies installed microwave radio links along these rights-of-way, particularly in the long, flat sections of the country or in the mountainous areas. Most railroad operators were more than willing to lease out use of their rights-of-way paths. It is probably a forgone conclusion that fiber-optic cable is or will be installed along railroad rights-of-way.

2.3.4.2 Leased microwave facilities. Some railroad operators had forethought and built their own microwave radio system along the railroad-controlled land. Towers were constructed along the side of the railroad tracks, and receive-transmit parabolic reflector antennas were mounted back-to-back. The transmit-receive equipment was either mounted on the tower, on a small equipment building, or on an enclosure at the base of the tower. These towers were installed at distances of approximately 30 miles apart.

One channel often carried diagnostic and monitoring signals for the microwave radio system. Analysis of this data could point to impending trouble so action could be taken before a major outage occurred. Heavy rains and lightning interference caused most of the outage or signal degradation problems.

The main purpose of installing such a system was to lease out microwave channels to the telephone companies. The only service the railroads needed was some small amount of railroad-scheduling data and the system-diagnostic monitoring information. Therefore, most of the channel communication capacity was leased out. Again, this provided good revenue for the railroad operators.

2.3.4.3 Telegraph/teletype communications. As mentioned before, the telegraph companies either leased a right-of-way for their own pole line or leased space for their wires on the railroad-owned poles. The telegraph industry grew from a manual key operation to a teletype system using TTY code. Telegraph and teletype systems were considered one of the first electrical digital wire communication systems. Today most people do not send telegrams, with the exception of banks, which use wire transfers of money. This system is in many cases not confined to a wire-type system but most likely uses several types of communication media, including fiber-optics.

3.1.1.1 Signaling, switching, and timing. The signaling, switching, and timing signals control the main connection that connects the calling party to the called party. If the telephone call is termed a toll call, then the length of time the connection is maintained governs the amount of the charges to the calling party. Therefore a timing signal or running clock has to be carried by the system. Often a separate circuit termed the common channel interoffice signaling (CCIS) system carries this type of information.

To operate the telephone set (sometimes referred to as the instrument or simply the phone), a direct current voltage source of –48 volts appears on the customer lines. To initiate a call, a caller takes the handset off the hook, causing the –48-volt source to form a current flow back to the local-exchange switching office. This current causes the switching system to find an open line. When the calling party dials a number, the switch bank goes through a row and column sequence to find the called party's line. If there is no loop current flowing in the called party's line, indicating the telephone is on the hook and not used, an 80-volt, 20-cycle alternating current source is initiated at the local exchange to cause the phone to ring. When the called party picks up the phone, loop current flows and allows the connection to be maintained as long as both telephones are off the hook and being operated. When one party hangs up, the connection is broken. If the called party is talking to someone else, loop current is flowing and the person trying to call that party will receive a busy signal triggered by the loop current flowing in the called party's telephone. This busy signal was caused by an electromechanical buzzer. Pulses of dual audio tones are used on the telephone system to advise the calling party of the status of the placed call. Table 3.1 describes these tone signals.

TABLE 3.1 Dual-Tone Telephone Signals

Tone signal	Dual-tones frequency Hz	On-time seconds	Off-time seconds
Tone	350 + 440	continuous	
	480 + 620	0.5	0.5
Ringback (normal)	440 + 480	2	4
Ringback (pbx)	440 + 480	2	3
Busy (toll)	480 + 620	0.2	0.3
Busy (local)	480 + 620	0.3	0.2
Off-hook	1400 + 2060 + 2450 + 2600	0.1 loud (0dBm)	0.1
Number	200 + 400	continuous modulated at A 1-Hz rate	

3

Cable System Transmission Methods and Controls

3.1 Hard-Wire Controls

Because copper conductive cable was used to connect the te
telegraph systems, a study of the types of signaling and contr
appropriate. Copper cables conduct electric current, and it i
this current that forms the message traffic. The telegrapr
make-and-break system to interrupt the current to form do'
Morse code. Later on, the invention of TTY code and the
ment to transmit and receive this code resulted in the tele
of the early submarine cables were operated in the telet
code. Messages were typed on a teletype machine, tran
and received by a machine that printed the message o

3.1.1 The telephone system

The present-day telephone system is a highly s
mit/receive system using copper wire, coaxial c
crowave radio systems as the communication r
signals found in typical telephone cable plant ?
voice signals making up the main message tra
timing signals necessary to make the proper

TA

dial t

busy

ring ba

ring bac

congesti

re order

receiver o

no such nu

The telephone system might seem to carry signals that might seem a bit antiquated. However, the telephone industry has never caused any of its equipment to become nonfunctional due to modernization of its plant, with possibly the exception of a nondial telephone set. The signals mentioned in Table 3.1 in most cases are generated electronically. Also, the switching banks in most local exchanges have been upgraded to solid-state switches, and the connection timing also is electronic. However, an old dial telephone set will still work.

Most present-day telephones use the touch-tone key pad. If a local exchange happens to still have an old electromechanical switch bank in operation, equipment at this exchange will translate the touch-tone information to the proper currents to operate the switch bank. Thus compatibility is maintained. The touch-tones generated by present-day telephones are dual tones similar to those listed in Table 3.1. However, the length is determined on how long each button is depressed. Figure 3.1 describes the touch-tone dialing method. Since the telephone rotary dial is quite failsafe, some protections have to be built into the dual-tone keypad so erroneous signals will not be sent throughout the telephone network. If two buttons in the same row or column are pressed, only one tone will be activated. If two buttons in a diagonal path are pressed, no tone will be activated.

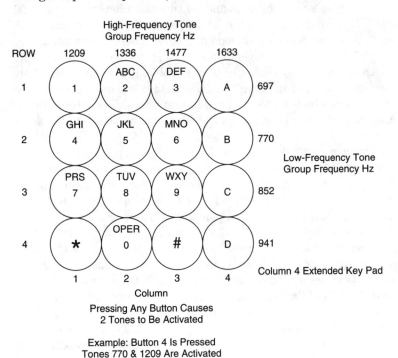

Figure 3.1 The DTMF (dual tone multiple frequency) telephone keypad.

When computer equipment is to be connected to the telephone system, an interface unit is needed. This interface equipment is called a modem and can be either a separate out-board piece of equipment or a printed circuit card that is installed in one of the computer's expansion slots. The modular telephone connecting wire and plug is used to connect the telephone system to the computer modem. The computer modem will react to the number it is programmed to answer. Therefore it responds to the ringing, supplies the loop current response to the calling party, and receives the message data and transfers it to the computer. If the computer is programmed to receive e-mail, the message will be appropriately stored in the computer memory for later viewing by the computer operator.

The computer modem contains all the necessary protocols used in the transferring of the received data to the computer. Modems are available that can take serial data from the telephone lines at the standard rates of 1200, 2400, 4800, 9600, up to 28,800 bps (bits per second). If a connection is made on the telephone lines and the data sent becomes garbled or full of errors, then the lines that connect the computers are inadequate to handle the chosen data rate. Another call should be initiated to connect the computers, and the line might then be good enough to transfer the data information correctly. After several tries and no success, a lower speed should be selected. Usually the 1200 and/or 2400 bps speeds are very successful. However, present-day high-speed modems contain a large amount of signal-processing and error-correcting circuitry. Therefore the telephone line quality does not cause any communications problems in most cases. Many businesses use computer modems to handle data transfers between office locations, many of which are located in different states. Computer information communications is a large part of the telephone industry's business.

3.1.1.2 Telephone network topology. The network topology of the telephone industry is quite unique and is referred to as a switched-star network. Beginning even with the local exchange office, the switched-star concept is used as shown in Figure 3.2. Local exchange offices are tied to higher-level switching exchanges via trunk lines. There are several classes of central offices designated as 1, 2, 3, 4, 4x, and 5. The 4x type is used for automatic digital services. The class 1 office is the highest class type, and there are only 10 or 12 in the United States and 2 in Canada. There are approximately 230 class 3 offices, 1300 class 4, and 19,000 class 5 offices in the United States and Canada. The switched-star network topology is carried beyond the local exchanges to various primary centers, toll centers, and sectional centers. Of course, long-distance service beyond the local exchanges constitutes a toll exchange. Figure 3.3 illustrates the switched-star network beyond the local exchange offices. Bidirectional, or what is called full-duplex telephone service, follows a pattern similar to the spokes of a bi-

For a Call to Originate at Telephone "A"
and Go to "B", the Path Has to Go through
the Local Telephone Company Exchange Switch.

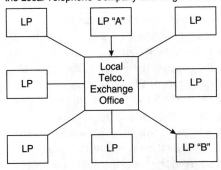

Figure 3.2 Switched star local exchange.

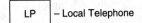 – Local Telephone

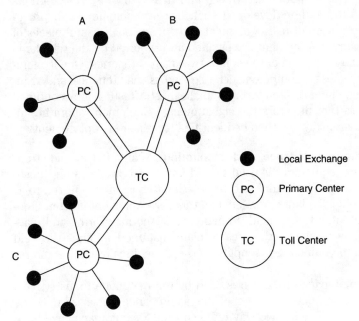

For Calling Areas A, B, and C, Connections from
One Calling Area to Another Are Switched and Controlled
by the Toll Center

Figure 3.3 Switched star networks connecting local exchanges to primary center
and primary center to primary center through the toll center.

cycle wheel. Offices and exchanges follow the same type of network config-uration known as the switched-star network topology.

3.1.1.3 System testing and measurement. It should be clear by now that the present-day telephone system is highly developed and very reliable. Most of us can seldom remember when we have had any serious telephone system failure. In order to achieve such system reliability, a rigorous testing program is used to constantly test various parts of the system automatically. Since the telephone communication network consists of a mix of copper wire pairs, coaxial cable, and fiber-optic cable, each type of cable system has to follow a different testing procedure employing different types of equipment. Testing is often conducted automatically using computer-con-trolled test equipment. Data is extracted from the measuring instruments in digital form, stored, and analyzed by the computer. The analyzed and stored data is used to activate an alarm when the data indicates an out-of-tolerance condition. The overall point-to-point signal testing consists of typical tests for signal-to-noise ratio, cross talk, echoes or reflections, and signal level.

Through its research branch, formerly known as Bell Labs, the telephone industry invented and/or developed such testing equipment and proce-dures. Manufacturers of test equipment make many such useful pieces of test and measuring equipment as specified and required by the telephone industry. Such test equipment falls into two main categories. One is moni-toring and diagnostic equipment. The second is maintenance and trou-bleshooting equipment. Monitoring and diagnostic testing is a continual testing process that monitors the signal quality. Equipment operating in-correctly or marginally can then be identified for repair or replacement.

3.1.1.4 System maintenance and monitoring. Maintenance and trou-bleshooting test equipment is used to find faults and problems on the so-called lines. Such maintenance and troubleshooting is often referred to as field work or outside plant work. Much of this test equipment is often time-battery-operated and hand-held. In many cases the monitoring and diag-nostic testing will indicate the area of plant malfunction. Thus the field work area can be pinpointed, simplifying the in-field maintenance and re-pair process.

The line maintenance procedures used by the telephone industry have been updated continually. Location of the maintenance trucks is known by the maintenance office by two-way radio control. System maps have been digitized and recorded on CD-ROM. A laptop computer system on the main-tenance truck can now present on-screen system map sections to aid in the troubleshooting and fault-finding procedure.

The telephone industry is very well known for its technical training pro-grams. Technical and maintenance personnel are required to participate in a continuing educational program to keep up with the new technologies and procedures. It is this highly trained technical staff that allows the telephone system to be so highly reliable. Since the software programs that operate most of the automatic switching exchanges are part of the overall reliability picture, proper software maintenance and diagnostics are also very important. A software bug could certainly create a problem if it is not discovered and corrected.

Maintenance of system performance records of past problems can be helpful in maintaining the telephone system. Such a history of problems and locations can be used to identifying areas for improvement and upgrades. Many of these records and histories have been digitized and stored in digital mass storage for quick retrieval and analysis.

3.2 Coaxial Cable Systems

As learned from previous chapters, coaxial cable has a much larger frequency bandwidth and hence a greater information transmission speed. With the development of radio-frequency communications, coaxial cable found use as a transmission line that was used to connect radio transmitters to antenna arrays. Research projects at the telephone company laboratories developed use of coaxial cable as a long-haul trunk-line application. Since the primary telephone business involves voice conversations between called parties, a method was developed to multiplex many voice channels on one coaxial cable. Most of the early research and development using coaxial cable was done by the telephone industry.

The cable television industry was developed around the coaxial-cable-connected repeater amplifier concept as investigated and developed by the telephone industry. The cable television industry started from humble beginnings using World War II surplus coaxial cable and often hand-built amplifiers. Present-day systems carry many television channels to customers homes, have an upstream capability, and are computer controlled. The cable television is a highly developed industry today and should be a major participant in the information highway.

3.2.1 Telephone system long haul

Since coaxial cable can carry electrical signals up to what is known as UHF (ultra high frequencies), a means of placing multiple telephone channels on an RF (radio frequency) carrier is required. The telephone industry devised a method using radio modulation techniques known as single-sideband suppressed carrier. The frequency bandwidth that carries the most intelligible parts of the human voice is the bandwidth of 3 kHz.

This is shown by the curve in Figure 3.4. The touch-tone dialing signals fall into the voice frequency band and are distinctly heard in the telephone

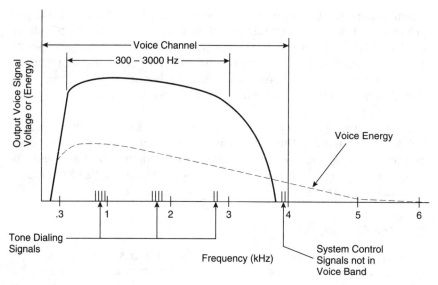

Figure 3.4 Voice channel band.

earpiece. The system control tones are still in the voice channel band but are attenuated to extremely low levels, so they are not heard. It is this voice channel band that has to be transmitted in both directions for a complete telephone conversation to take place.

3.2.1.1 Voice channeling/multiplexing. The method devised by the telephone industry to stack voice channels on a coaxial cable system uses the single-side-band suppressed carrier (SSB) technique used by the radio communication industry and the amateur radio operators commonly referred to as (HAMs). This method takes each voice channel, mixes it with a selected carrier frequency, and selects the sum frequency band. If a different carrier frequency is selected for each voice channel, the sum frequency bands can then be combined into a group of contiguous bands. This method is shown in Figure 3.5.

This signal is either transmitted along a coaxial cable system to its destination or combined with other 12-channel groups to form 24, 36, and 48 voice channel groups. Since this plan was worked out by the telephone system in the United States, it became accepted as an international standard and hence has an international designation. The international group that was formed to assure compatibility between telephone systems is known as CCITT (Telephone Telegraph International Consultative Committee). The method shown in Figure 3.5 illustrates the principal only. This technique has been updated and changed, and a typical channel plan is shown in Figure 3.6 for the formation of a CCITT standard group of 12 voice channels

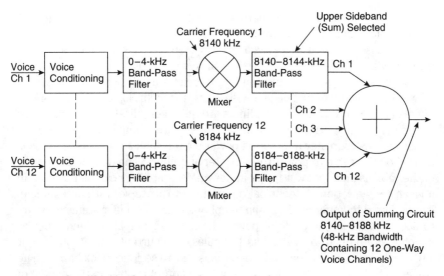

Figure 3.5 Coaxial cable 12-voice channel carrier method.

in a single direction. Another similar system has to be used for the other half of the telephone conversation in the opposite direction. The same technique is used to form supergroups made from 5 standard groups giving 60 voice channels. The formation of a master group consists of 5 supergroups

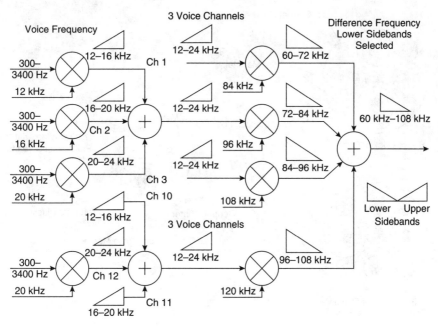

Figure 3.6 A method for forming a CCITT group of voice frequency telephone channels.

for 300 voice channels. The supermaster group consists of 3 master groups for 900 voice channels in a given direction.

As discussed previously, a coaxial cable system has loss that increases with frequency and distance. The cable loss at lower frequencies is less than the loss at higher frequencies, and the loss increases in dB per 100 feet or 100 kilometers as a function of distance. To make up for this loss, a cascade of amplifiers are employed. These amplifiers are quite complicated, and they have automatic gain control (AGC) and temperature compensation circuitry. Usually a low-frequency alternating power source such as 60 volts at 60 Hz is carried on the cable to power the repeater amplifiers. The dc loop resistance of the cable is an important factor in how many amplifiers can be powered in a given cable section. A technical discussion of this topic, with an example, can be found in appendix D. Some telephone coaxial cable systems carry up to 650 volts dc on the cable to power the repeater amplifiers, and a power supply section might be as long as 160 miles. The current is limited to about 110 mA, but this level of 650 volts does constitute a serious shock hazard. The placement of cable power supplies should be approximately in the middle of the cable span.

Coaxial cable systems used by the telephone industry are at a nominal characteristic impedance of 75 ohms. Several coaxial cables appear in a common outer jacket, and filler material is usually placed in the spaces between the cable bundle. Figure 3.7 shows a few typical coaxial cables.

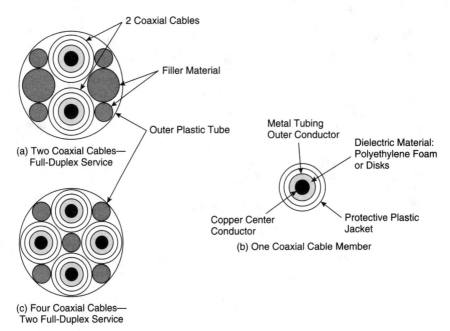

Figure 3.7 Typical telephone coaxial configurations.

Remember that it takes two coaxial cables to provide full-duplex service. Usually a pair of cables that form duplex service are placed in the same tube. The telephone industry has classified its cable transmission systems according to the number of voice channels carried on a single cable. The greater the number of channels, the higher the line frequencies become. Hence, more repeater amplifiers in cascade per span of cable are required. Table 3.2 give the characteristic for L-type classes of cable systems. The L_1 and L_3 systems have the capability of carrying video channels for NTSC (National Television Study Committee) type video signals. An example of a coaxial cable transmission system in a full-duplex configuration is diagrammed in Figure 3.8.

3.2.1.2 Bridging between system networks. Telephone coaxial cable systems connect areas together, acting as trunk lines. The coaxial cable carriers have to be demodulated and the telephone conversation signals switched to hard wire to operate the telephone sets and computer modems. This operation takes place at repeater terminals where service adds or drops to the telephone wire circuits are made. Routing of the telephone connections through the various available trunking methods are accomplished automatically by computer-controlled switches. Also, the routing and time of connection determines the cost to the user.

TABLE 3.2 Characteristics of Telephone L-Type Coaxial Cable System

Name	L system L_1	Designation L_3	L_4	L_5
Maximum length of cable span in miles	4000	1000	4000	40000
Number of 4 kHz voice channels (FDM)*	60–2788	312–8284	864–17,548	1590–10,800#
Line frequencies kHz	60–2788	312–8284	564–17,548	1590–68,780
Repeat amplifier spacing miles	8	4	2	1
Power insertion points miles	160 or 20 amplifiers	160 or 42 amplifiers	160 or 80 amplifiers	75 or 75 amplifiers

*Frequency Division Multiplexing   #Specification L5E 13,200 channels

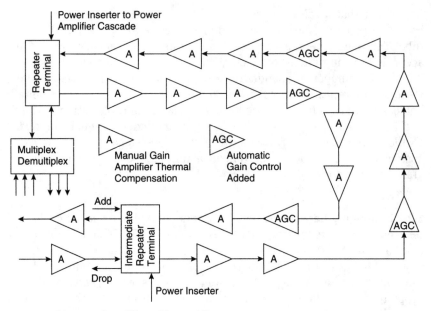

Figure 3.8 Diagram of amplifier cable cascade.

The carrier channel plan for telephone coaxial cable systems was extremely well thought out. If the route of the service involved areas where there was no coaxial cable, a microwave radio link could bridge cable sections directly. Essentially the whole group of line frequencies is transferred to a microwave carrier and transmitted to both sections of a coaxial cable system. At the receiving end, the microwave carrier is demodulated and the resulting coaxial cable signal is applied to the next coaxial cable section. This technique is shown in Figure 3.9. So far, the methods of transmission for voice traffic have been analog. Electrical power at voice frequencies is analog, so the development of analog transmission methods was natural. With the development of the digital computer, the transfer from the analog domain to the digital domain for many analog signals was required. Digitization of voice frequencies and hence telephone communications was quickly developed, and many of the existing coaxial cable systems and microwave links were later converted to digital signal methods. Various systems of signals and transmissions are taken up in chapter 4.

3.2.2 Cable television systems

Cable television systems are basically coaxial cable systems that deliver television programming to subscribers' homes via a cable-connected cascade of repeater amplifiers. In chapter 1, Figure 1.3 shows an elementary cable television system. Present-day cable television systems, often referred to as CATV systems, appear in many network configurations. Early cable television systems consisted of one cable from the receiving antenna site to the subscriber area. The television signal was tapped off this cable and fed to the subscriber's home via a drop cable connected to the subscriber's television set. When the signal quality became bad enough from the noise and distortion buildup caused by the repeater amplifiers, the service area was terminated.

A major improvement in plant design, called the *trunk feeder concept*, allowed the service area to be further expanded in size. Trunk lines carried the signal from the central receiving antenna site (called the head-end) to the edges of the system service area. These trunk cables carried the signal, and no subscriber taps were connected to the trunk. At repeater amplifier locations, a bridging isolation amplifier took a sample of the pure trunk signal, increased the signal level to its proper value, and fed separate distribution cables containing the subscriber taps. These distribution cables would only allow two or three distribution repeater amplifiers. For the case of a service area that occurs between normal trunk amplifier locations, an intermediate bridging amplifier is required to drive the distribution cable serving that area. This method is referred to as a *tree/branch system*. The trunk system acting as the tree trunk delivering the signal to the branches and twigs.

3.2.2.1 Cable television coaxial cable plant. The tree/branch architecture became the bread and butter of the cable television industry. Since the central receiving site on head-end represented a large portion of the cost of a cable television system, methods were devised to feed several cable systems from a single head-end. Three methods were developed that performed this function. One method is a microwave radio, another is

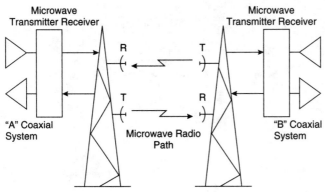

Figure 3.9 Telephone microwave radio coaxial cable system "A" and coaxial cable system "B" together.

supertrunking, and the third is fiber-optic cable. The most recent method developed is the fiber-optic cable system. Still, all have their good and bad points. Microwave radio could transmit the whole television spectrum a distance of approximately 30 miles line-of-sight to remote areas requiring no utility-pole-mounted cable system. This technique is shown in Figure 3.10. The master receiving head-end provides cable systems A, B, and C with service via microwave radio and cable system D directly by cable. Often the microwave link would only distribute certain signals not obtainable separately by each cable system. With the sharp decrease in the cost of satellite receiving systems, it was often less costly for each cable system to have its own satellite receiving system. Now only the hard-to-receive channels were transmitted via microwave radio to the various remote cable systems.

Where there were pole lines connecting the remote cable systems to the central master head-end, the choice of supertrunking was available. For the coaxial cable method of supertrunking, large-size coaxial cable is often chosen because of the less loss per 100 feet or 100 kilometer specification. Since loss is also less at lower frequencies, the selected channels were often

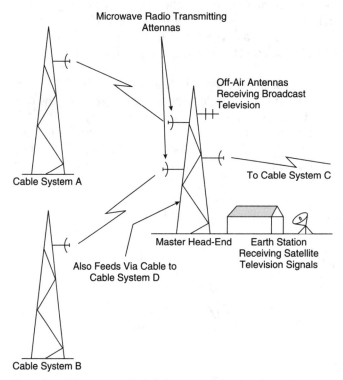

Figure 3.10 Microwave radio link from master head-end to cable system "A," "B," and "C."

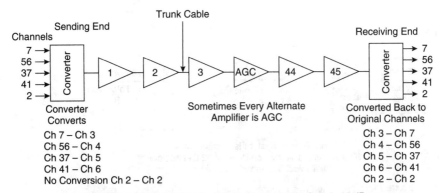

Example: A Cascade of 45 Amplifiers. The Amplifier Gain Is 22 dB
and Cable Span Loss between Amplifiers Is 22 dB. For a Cable with a Loss of 0.32 dB/100'
at Ch 6, the Span Distance is: 22 dB/0.32 dB per 100' = 68.75 × 100' = 6875 Ft between
Amplifiers. For 45 Amplifiers, the Distance between the Sending End and the Receiving End is
45 × 6875 ft = 309375 Ft. In Miles, 309375/5280 = 59 Miles. This Distance of 59 Miles, If
Done with a Microwave System, Would Need Two 30-Mile Hops.

Figure 3.11 Supertrunk cable cascade.

converted to lower frequencies. Now the repeater trunk amplifiers could be spaced farther apart. Also, amplifier input filtering allowing only the signal passage helped limit the broadband noise buildup.

The coaxial cable supertrunk concept is shown in Figure 3.11. Supertrunking of cable signals is considered an expensive method that requires a whole cable system to connect the master head-end source to a remote cable television system. Extra power is consumed, and the system has to be maintained to high standards to assure good signal quality to the remote site. Microwave radio is also considered expensive, so many factors have to be considered when facing this choice. Often the choice is clear because some remote cable systems might not have a utility pole line, and microwave is the only choice.

The alternative to coaxial cable supertrunking is a fiber-optic cable interconnect. This method requires a fiber-optic cable to connect the sending station to the receiving station. A fiber-optic transmitter at the sending end converts the television band to optical energy and into the fiber-optic cable. At the receiving station, an optical receiver converts the received light energy back to the radio frequency television band. At this receiving end, the signals are amplified and combined with any required local signals to feed the local cable system. Figure 3.12 shows this technique.

This method is quite efficient because it can relay a television band, for example, of 62 television channels a distance of approximately 25 miles in a single hop. If further distances are required, another repeater site can go another 25 miles. If fewer channels are required, some increases in dis-

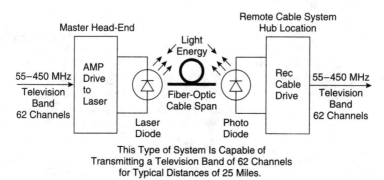

Figure 3.12   Fiber-optic feed to a remote cable system.

tances between stations can be realized. Since fiber-optic systems can relay television bands of frequencies so well, system designers are employing fiber-optic methods into the overall cable television designs. Essentially, fiber-optic transmission methods are replacing the coaxial trunk cables. Designers are feeding distribution modes with pole- or pedestal-mounted receivers, which in turn drive the area with traditional cable plant technology. This concept eliminates the trunk cable cascades, which in turn saves electrical power and maintenance costs. Figure 3.13 illustrates this method, which is often referred to as the fiber-optic backbone. Many cable television operators are redesigning their cable systems, eliminating long trunk amplifier cascades, which in turn increases signal reliability. There is less equipment to malfunction. Since the long amplifier cascades affect the number of television channels carried, going to a fiber-optic backbone technique will often allow more channels to be offered. Fiber-optic methods have definitely offered cable operators a large benefit to their businesses.

Remember that the network architecture discussed so far has been the tree-branch, one-way or downstream delivery of television service to subscribers. Early in the development of cable television, there was considerable interest in some upstream or reverse system programming. Such programming as public school educational media or municipal programming originating at the various school and/or municipal buildings could travel upstream to the central head-end or hub station to be distributed downstream to the subscribers. Then school or municipal programs could be viewed by the subscribers. Most cable television operators were required by the municipal licensing authority to maintain some local facilities for the production of local programming. Although local programming and reverse-system capability was an added expense to the cable operators, many cable systems regarded this added programming as a benefit. Only cable sub-

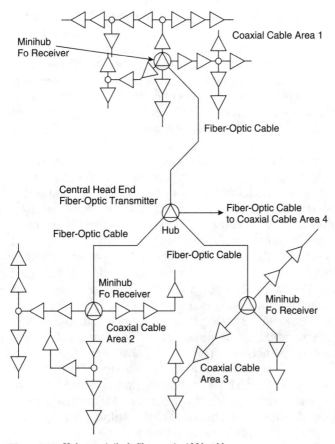

Coaxial Cable Area 1

Minihub
Fo Receiver

Fiber-Optic Cable

Central Head End
Fiber-Optic Transmitter

Fiber-Optic Cable
to Coaxial Cable Area 4

Fiber-Optic Cable   Hub

Fiber-Optic Cable

Minihub
Fo Receiver

Minihub
Fo Receiver

Coaxial Cable
Area 2

Coaxial Cable
Area 3

Figure 3.13 Hub-to-minihub fiber-optic AM backbone.

scribers could view these services making this an inducement to subscribe to the cable television service.

A method for using the same cable for upstream and downstream transmission was devised. Essentially, the frequency bandwidth was divided, with a band of frequencies allocated as downstream and a band for upstream. These bands were described in chapter 1. The success of using the reverse single cable method was plagued by noise, basically because of the network topology. This noise buildup was referred by the cable television industry as the reverse tree buildup. As mentioned in chapter 1, the frequency bands were split into what is known as subsplit, midsplit and highsplit. The subsplit only offered 4½ reverse channels operating in the 5–30 MHz band. The space between 30 MHz and 55 MHz (TV channel 2) was unusable and separated by a crossover filter separating the upstream and downstream services. In many cases, the four channels were adequate.

The reverse-tree noise buildup is illustrated in appendix E. The accumulated noise traveling along the branches back through the trunk to the head-end hub site depended on the number of branches. If the area to originate a program to be sent upstream was geographically near the head end, the branches preceding this area could be disconnected for upstream service. This cut off the areas that were contributing noise and allowed a clean, noise-free signal arriving at the head-end. At the head-end site, the program video/audio signal was detected and retransmitted down the forward system on one of the standard television channels.

Because some plant arrangement was necessary in order to use the reverse system, the concept of automatic bridger-leg switching was developed. When a reverse signal channel was to be activated, the bridger-leg switching system would disconnect and terminate the not-needed sections of reverse feeder plant, thus allowing only that section of needed upstream feeder to be active. This technique significantly reduced the buildup of noise. Computer control of these remote switches was developed so different reverse sections could be activated at the hub head-end location. Digital data was sent on a downstream frequency channel, often in the upper FM frequency band, which activated the desired return path.

Since this type of system cost was substantial, some cable operators elected to install a separate upstream cable system, sometimes referred to as an I-net or institutional network. This type of plant was often a barebones type of system because only a few pickup points along the path were necessary. An example of such an I-net or shadow trunk is shown in Figure 3.14. This technique provides a pure television signal deep in the system, thus eliminating significant portions of coaxial cable trunk. Many systems kept the coaxial trunk plant in place and could use it in the event the fiberoptic plant failed. Some innovative cable operators converted some trunk sections to feeder plant to supply new or built-up areas with cable television service. An example of such a conversion is shown in Figure 3.15.

Although the capability of reverse or upstream service for a cable television system remains, many cable television operators do not use and have not activated the reverse system. Thus, many typical cable systems are one-way, downstream, tree-and-branch network architectures. This downstream plant has typical downstream frequency bandwidths with an upper frequency limit of 300, 400, 450, 600, 750, or 1000 MHz, depending on the system design parameters.

3.2.2.2 Signal measurement and testing. Cable television system testing and measurements should be performed to assure that the cable plant is operating as per specification and that high-quality signals are being delivered to subscribers. In order to test any system or cable plant, a thorough understanding of the principles of operation should be learned.

Cable television technology is based on the unity gain cable-amplifier sec-

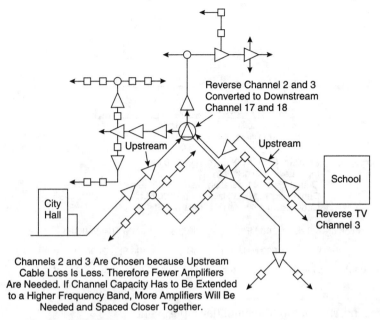

Reverse Channel 2 and 3
Converted to Downstream
Channel 17 and 18

Upstream

Upstream

School

City
Hall

Reverse TV
Channel 3

Channels 2 and 3 Are Chosen because Upstream
Cable Loss Is Less. Therefore Fewer Amplifiers
Are Needed. If Channel Capacity Has to Be Extended
to a Higher Frequency Band, More Amplifiers Will Be
Needed and Spaced Closer Together.

Figure 3.14 I-net reverse cable system.

tion. Much of the early work was done by the telephone industry in design-ing its bidirectional amplifier coaxial cable systems used for long-haul tele-phone trunking. This design principle dictates that each coaxial cable section has the same amount of loss that the amplifier has gain at the high-est operational frequency.

If the highest frequency channel is increased to a higher frequency, then the cable loss is greater, and more amplifiers for a given distance are re-quired. This principle is shown in Figure 3.16. As might be suspected since the input signal to an amplifier is designed to be correct at the highest op-erating frequency, it would be too high at the lowest operating frequency, usually TV channel 2 (55-MHz video carrier frequency). In order to prevent the amplifier from being overdriven at the lowest frequency, an equalizing

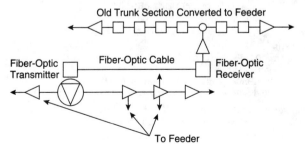

Old Trunk Section Converted to Feeder

Fiber-Optic
Transmitter

Fiber-Optic Cable

Fiber-Optic
Receiver

To Feeder

Figure 3.15 Use of existing trunk cable converted to feeder system.

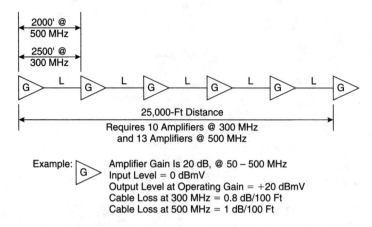

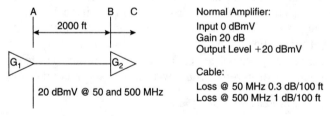

Figure 3.16 Amplifier-cable distance as determined by highest operating frequency.

network is inserted before the amplifier input point.

This network is a frequency-dependent network that has more loss at the lowest frequency and negligible loss at the highest frequency. The equalization method is shown in Figure 3.17. In practice, designing equalizers that have this much loss at low frequencies and zero loss at high frequencies is difficult. The technique of tilting the amplifier's output signal at high frequencies means that the signal level at the downstream amplifiers' input will be larger. Therefore the equalizer will not need such a large slope.

Figure 3.17 Cable equalization.

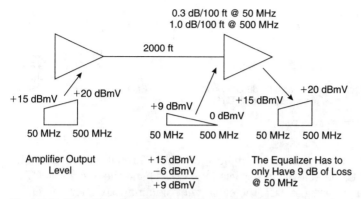

Figure 3.18 Block tilt method of cable section equalization.

Figure 3.18 illustrates this compromising method. This is often referred to as the *block tilt technique* because the block of frequencies is tilted upward at the amplifier output to help compensate for the downward-tilted cable loss. Now the amount of the equalizer tilt is not as great.

The trunking system used in many present-day typical cable television systems carries the signal to the ends of the system. The lengths of these trunk cascades are limited by the number of amplifiers allowed in the cascade before noise and signal distortion build up sufficiently to impair the signal. If allowed to become too large, the noise and distortion components will cause the received television pictures to become poor. The trunk amplifier cascade of cable sections and repeater amplifiers is similar to one-half of the duplex cable cascades used by the telephone industry. Since the output of one amplifier is connected through a cable section to the input of the next amplifier, some of the noise generated by the first amplifier is going to be amplified by the succeeding downstream amplifiers.

The first thing to do is establish the noise base value and then apply this to the amplifier cascade. The noise floor and the buildup of noise in amplifier cascades is discussed in appendix F, along with the buildup of second- and third-order amplifier distortions. Since the noise power doubles as the number of amplifiers doubles, this doubling of a power level corresponds to an increase of 3 dB. Therefore if the noise level increases by 3 dB and the signal level at the amplifiers output remains constant, then the signal-to-noise ratio expressed logarithmically in decibels means that the signal-to-noise ratio decreases by 3 dB for every doubling in amplifier number, as shown in Figure 3.19. Second-order distortion builds up in the same manner. Third-order distortion increases on a voltage basis or 20 times the logarithm of n the cascade number.

Testing procedures for noise and distortion measurement for cable television systems has been well established. Early work on testing has ap-

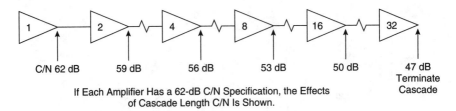

C/N 62 dB   59 dB   56 dB   53 dB   50 dB   47 dB
                                                      Terminate
If Each Amplifier Has a 62-dB C/N Specification, the Effects   Cascade
of Cascade Length C/N Is Shown.

Figure 3.19 32 Amplifier cascade.

peared in telephone industry technical journals. Through efforts made by the National Cable Television Association (NCTA) and the Society of Cable Television Engineers (SCTE), the cable television industry has been effective in improving and promoting proper measurement procedures and establishing standards and specifications for system performance.

The NCTA has published a manual of measurement procedures and methods and has made it available to the industry. The SCTE has established a technical training system consisting of area training seminars, books, videotapes, and satellite-transmitted technical lectures that have effectively provided most of the available technical training. Some of the larger cable television operators have instituted in-house training schools for their company technicians. One company even made its company school available to other companies' people at a nominal cost.

Unfortunately, the cable television industry in general has not had the intensive technical training programs provided by the telephone industry. The SCTE has established a certification program that provides the cable television industry with technical training standards for system technical personnel. Levels of expertise are at the installer technician level, system and service technician levels, and engineer level, which is obtainable by examination only. For the most part, these standards are used by many cable television operators. Other training programs are by correspondence schooling through the National Cable Television Institute (NCTI) or other correspondence schools. In general, formal school courses in cable television are scarce. Testing and measurement procedures for carrier-to-noise ratio, carrier-to-low-frequency noise and hum ratio, and carrier-to-distortion ratio are discussed in appendix G.

These tests are made on or in relation to the television video carrier frequencies, not the video baseband signal itself. Distortion beats and carrier-to-noise ratio indeed do affect the video signal quality, and not until recently has baseband video signal quality testing become a requirement. Testing of cable television plants fall into two main areas. The cable system has to be tightly sealed so as to not leak signals into the surrounding area affecting normally licensed television broadcast and radio communications services. In its rules and regulations for cable television systems, the

Federal Communications Commission (FCC) has specified the minimum leakage specifications and requires cable television systems to test their systems on a continual basis and file their reports once a year. This report proves to the FCC that the cable system conforms to the allowed leakage standards. The specification acceptable to the FCC is the Cumulative Leak Index (CLI), which is calculated by the following formula:

$$CLI = 10\log \left[\frac{\text{total plant miles}}{\text{tested plant miles}} \times \text{sum of (leak)}^2 \right]$$

Each leak is converted from the dBmV leak reading to microvolts per meter at the test frequency.

Since many cable companies carry television channels almost completely across the available cable bandwidth, sensitive over-the-air communications services such as commercial aviation safety and air navigation aids could be affected by the buildup coming from leaking cable systems. This is why the FCC considers this compliance to leakage standards so important. Cable television systems failing to conform are subject to heavy fines. The FCC also maintains its own field test and mobile monitoring system to sample-test cable systems around the country. Through the Amateur Radio Relay League (ARRL), the amateur radio operators have also performed leakage monitoring and have in many cases aided cable television operators in tracking down leaking cable sections. Many test-equipment manufacturers have introduced accurate and innovative leakage test equipment. The leakage testing procedure is discussed in appendix H.

The second area of testing is the previously discussed noise and distortion testing, with the additional tests on signal level and frequency. This is often referred to as "proof-of-performance" testing, which is required to be done twice a year by the FCC. The cable operator has to have it on file at the company office and ready for inspection, provided a representative of the FCC arrives and announces an inspection. The testing of frequency is usually performed at the head-end of the system, and the noise and distortion system tests are taken at an amplifier location at the extremities of the system, one of which has to be at the longest amplifier cascade. The system level test is also usually conducted deep in the system cascade so as to represent a so-called worst-case condition. The sound carrier is usually set to 10 to 17 dB below the video carrier level, which reduces adjacent video channel interference caused by the lower channel audio signal showing up in the upper adjacent channel picture.

For some systems with long amplifier cascades, the television channel frequencies can be shifted or offset to reduce the visible affects of third-order distortion. One channel plan is termed Incrementally Regulated Carriers (IRC), where TV channel 4+ appears in between normal channels 4 and 5. TV channel video carriers 5 and 6 are increased to 79.25 MHz and

85.25 MHz to help create the space required by channel 4+. For this channel plan, all television channels are spaced 6 MHz apart, beginning with channel 2 at 55 MHz. A precision generator driving the signal processors at the cable head-end keeps the channel frequencies in proper alignment.

Another channel plan begins at 5 MHz and spaces all of the television channels at a precise 6 MHz spacing. This system uses a precision comb generator to phase-frequency lock all of the head-end processors. This channel plan is referred to as harmonically regulated carriers (HRC). The IRC plan is very close to the standard plan, and the HRC system is quite different. This causes problems with normal television channel selection. Cable operators have available set-top converters that perform the channel selection and convert any of the selected channels to one channel (3 or 4) selected by the TV set. Many present-day cable-ready television sets have a selector switch allowing for normal, cable 1 or cable 2 positions that correspond to antenna, IRC, or HRC channel alignment. The IRC plan allows the third-order distortion to be 4–6 dB worse than normal before any visible picture degrading is seen. HRC allows 6–9 dB worse than normal third-order distortion. For frequency testing, the controlling generator should be tested first before each channel is measured.

Frequency testing and measurement are discussed in appendix I. This is an important parameter that effects picture quality. This test is conducted at the cable system head-end where the frequency determining equipment is located. For consecutive television channel operation, the video and audio carriers have to maintain precise spacing between adjacent television channels. Also, the level of the audio carrier is set between 10 and 17 dBmV down (lower) from the video carrier level so as to not interfere with the upper adjacent video carriers spaced up by 1.5 MHz.

3.2.2.3 Cable television standards and specifications. Cable television standards and specifications are regulated by the FCC Rules and Regulations part 76. The first writing of part 76 appeared in 1972 and only applied to the so-called class 1 television broadcast stations. During this time, the cable television industry was starting to develop and was still known as CATV (community antenna television). This was an appropriate name because a cable television system merely acted as a large supertelevision antenna to subscribers. The early rules set the standards to protect the television broadcast stations' signals and assure subscribers proper service. Still, cable operators were allowed to set their own standards for nonbroadcast signals such as satellite and/or locally generated channels.

In the past, the FCC has relaxed and tightened the standards as the industry has developed. In the early 1980s, the problem of signal leakage from cable systems became a problem mainly because many cable systems were formed that were carrying cable channels in the frequency bands occupied by aircraft emergency and navigation services. Cable operators thought their systems were sealed, and they were allowed to use the whole cable bandwidth. The accumulated leakage from cable systems can, if allowed to occur, affect other licensed services. The amateur operators have reported leakage occurring that caused interference. Most cable operators have successfully maintained their plant from causing problems and have worked with the Amateur Radio Operators and local authorities.

As discussed in the previous section, some cable systems offset the frequency channel alignment according to IRC and HRC specifications. Also, television channels occurring near an aircraft frequency near an airport could be offset away from the critical aircraft frequency. Therefore if some leakage occurred, it would not fall on the critical frequency. In 1985 the FCC did not require cable operators to test their systems for signal quality as per part 76. However, the signal leakage testing and maintenance program instituted earlier was left intact. At this time the commission must have felt that the marketplace could determine signal quality. In 1992 the commission reestablished the standards and also added some standards with the agreement and encouragement of the cable television operators and the local regulating authorities. These are the rules that apply to today's cable operators.

The FCC specifications and technical standards are divided into two main categories. The first category is the requirements pertaining to the radio frequency video and audio carriers that are carried on the amplifier cascade. The second category pertains to the video baseband signal distortions. For the RF television carriers, the parameters and values will be listed and discussed.

Audio carrier. The frequency of the audio carrier will be maintained 4.5 MHz ±5 kHz above the video carrier frequency for the television channel under consideration. This means that a frequency counter will have to measure the frequency to 4.505 MHz upper limit to 4.495 MHz lower limit. The audio carrier level is maintained between 10 and 17 dB below the channel's video carrier level.

Video carrier. The signal level at a subscriber location will be at least +3 dBmV at the end of a 100-foot drop from the subscriber tap port. The signal level must be at least 0 dBmV at the subscriber's television antenna terminals. Variation in video signal level during a 24-hour period in July or August (summer) and a 24-hour period in January or February (winter) cannot vary for each channel more than 8 dB over any 6-month period.

Amplitude characteristic. Over the bandwidth of any one specified television channel, the response will not vary more than ±2 dB from 0.75 MHz to 5 MHz above the lower channel boundary limit measured at the subscriber terminals. The characteristic is illustrated in Figure 3.20.

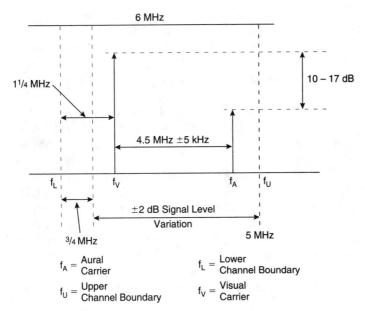

Figure 3.20 Television channel response.

Carrier-to-noise ratio. The minimum allowed carrier-to-noise ratio is 36 dB as of 4/92, and in one year, 4/93, the carrier-to-noise ratio increased to 40 dB. Within the three-year period, the minimum carrier-to-noise ratio will be increased to 43 dB. This parameter is measured at the subscriber's terminals.

Signal level to coherent disturbance. This is the distortion specification for any type of amplifier distortion, which includes second and third order, intermodulation, and any discrete frequency interference. This specification is at 51 dB video carrier level to distortion level for noncoherent channel alignment systems. For coherent systems (IRC/HRC), the specification is relaxed to 47 dB. This measurement is allowed to be measured using modulated carriers and averaged over a sufficient time period.

Terminal isolation. This parameter means that sufficient isolation between adjoining subscribers must be maintained so as they will not interfere with each other. This isolation is usually meant to be between ports on a given tap. Present-day taps are well-enough designed and manufactured, so this is not a problem. The commission allows the use of manufacturer's specification for port-to-port isolation to be sufficient evidence of proper subscriber isolation.

Hum and low-frequency disturbances. This parameter is a measurement of the signal (video) level versus the level of hum, ripple, or any low-frequency disturbances (usually related to commercial power-line frequency) in percent. The maximum allowed is 3% of the video signal that is hum related. This parameter used to be 5% but was tightened to 3%.

The next parameter requiring testing is the measurement of the NTSC video baseband color-television signal and is usually conducted at the system head-end.

Differential gain. This is a measure of the difference in amplitude between the greatest and smallest part of the chrominance (color) signal, divided by the largest component and expressed as a percent. The specification allows 20 percent. Most cable television systems are significantly better than this number and are usually around 2–3 percent. Several manufacturers offer instruments that allow this test to be made easily.

Differential phase. This parameter is the difference in degrees between each part of the chrominance (color) signal and the reference part (at the blanking video level). This difference cannot be over $\pm 10^0$. The test procedure for differential phase can be made using a vector scope or a specially fitted waveform monitor with the necessary signal filters. There are now some available instruments that make this test easier.

Chrominance-luminance delay. This delay specification is the delay in nanoseconds (ns) of the chrominance signal to the luminance (brightness) component and should not be greater than 170 ns. This parameter is normally not greater than 45 ns for most systems.

These three video tests by rights should be conducted on all baseband-controlled cable television video signals. Most cable television systems normally use signal processors for off-air broadcast channels and hence do not control any of the baseband video signals. Usually, locally generated or re-modulated baseband signals should be quality tested for these parameters.

Cable television systems with less than 1000 subscribers are exempted from FCC testing. The commission also forbids local authorities from requiring cable operators to conform to standards more stringent than the FCC standards. It also does not require cable systems to improve the signal quality of its received (off-air, satellite, etc.) signals. The commission forbids systems to alter or remove the closed-captioned information supplied for the hearing-impaired subscribers. This information is supplied on line 21 or the VBI portion of the video signal.

The proof-of-performance tests for the RF television carriers are required to be conducted semiannually on four television channels, with one channel added for each added 100-MHz increment of system bandwidth. A 216 (21-channel) system would only have to test on 5 different channels for the first 100 MHz, and 1 more for the next 100 MHz. For 400 and 450 MHz, testing will be made on 4 channels, plus 1 for each added 100 MHz—7 channels at 400 MHz and 8 for 450 MHz.

Since cable television systems follow the tree-and-branch network architecture, these tests are required to be made on a minimum of six points in a given cable-connected service area. A test point must be added for each 12,500 subscriber increment. The three baseband color-television signal tests only have to be conducted (at the head-end) every three years. These specified parameters on the color-television signal as required by the commission are not stringent or difficult to attain with normal industry equipment. The tests required by the commission assure that systems maintain proper operating standards and keep such records as necessary for FCC inspection.

3.3 Other Commercial Coaxial Cable Systems

Another type of cable system that has had remarkable growth is the computer system interconnect often referred to as a local area network (LAN). When computing equipment in the form of the personal computer became a standard piece of office equipment, the need to interconnect office to office became quickly apparent. Thus some method was needed to connect the parallel or serial input/output ports from one computer to another. Large companies operating in a campus environment required digital information to be transferred between offices in separate buildings. The LAN system at first was contained in a single building and expanded to a building-to-building, office-to-office configuration.

The computer manufacturers were the first to realize the need to interconnect computer equipment. IBM, Digital Equipment (DEC), and Apple Computer all contributed to developing various interconnect plans. The Institute of Electrical Electronic Engineers (IEEE) formed a committee to study various proposals and established some standards before the situation became worse. Industrial/commercial need provides the fuel and spark for new product development, therefore several methods of LAN designs were developed, sold, and installed. Many of the earlier-designed LANs are still successfully operating today.

As the size of some LAN systems grew, the term wide area network (WAN) appeared. As a rule-of-thumb, a LAN spans a distance of approximately 10 km. Greater distances would therefore be classified as WANs. Some networks operated by municipalities that connected various offices and buildings within a city's limits were referred to a metropolitan area networks (MANs).

Data rates for such networks as LANS/WANS are somewhat different, mainly because for the shorter-distant LANs, the data-rate requirement might only need an upper limit of 10 million bits per second (Mbps). The data rate used by a WAN could be quite slow if the input/output connection is a dial-up or leased telephone line, which is typically up to 28.8 kilobits per second (Kbps) with present-day telephone modems. Where optical fibers connect the buildings together, the bit rates typically become 100 Mbps.

The typical type of equipment connected by an LAN/WAN communication system would be computer to computer terminal (workstation to workstation), computer terminal to mass storage (file server), and computer terminal to printer/production location. Data can then be requested or delivered to various pieces of equipment, thus streamlining office procedures. LAN/WAN installation has had a great cost-saving effect on the way business is done.

3.3.1 LANs—Local Area Networks

As might be suspected, simply placing plugs and connectors in equipment is not the way a LAN system is installed. Computer programs have to have instructions to activate their input-output ports to transfer data. Therefore some form of interface system had to be developed to do this job. A network interface unit (NIU) in the form of a built-in printed circuit card or an out board (separate) box that connects the LAN to the computer workstation is required for LAN/WAN operation. This NIU separates the transmission method of the LAN to digital computer methods of operation. Like any proper transmission system, it should be essentially transparent to the information message.

The transmission medium of typical LAN implementations are of three types. For LANs covering small areas that do not require high data rates, twisted-pair standard-telephone-type cable is used. Since all the connectors, tools, and instruments developed and used by the telephone industry can be used, twisted-pair technology offers a distinct economic advantage to this type of system. For higher data transfer rates, coaxial cable is often used. This cable can support a high data bit rate per cable for a baseband digital system. Coaxial cable can also use CATV technology and carry a band of carriers upstream and a band downstream with digital modulation in a bidirectional configuration. For many later-constructed LANs, fiber-optic cable is often used. This cable is immune to electrical noise and has such low values of optical loss that it needs almost no signal-repeating equipment. Fiber-optic technology offers extremely high data rates, which permit an enormous amount of data to be transferred in a short amount of time. As might be suspected, there are advantages and disadvantages to each of these cable systems, which will be later studied and analyzed.

3.3.1.1 Types of LAN systems. As stated earlier, several of the computer manufacturers foresaw the need to interconnect their equipment, so they offered several types of cable for such purposes. IBM offered four types of cables, referred to as types 1, 2, 5, and 6. Each cable is used for different purposes, and these are summarized in Table 3.3. Digital Equipment Corporation (DEC) developed a line of cables referred to as DEC connect, which is listed and discussed in Table 3.4.

TABLE 3.3 IBM-Type LAN Cables

Cable type	Description	Recommended uses
1	Two # 22-AWG twisted-pair solid copper wires with braided-copper-tinned shield with insulating plastic jacket.	Available for indoor and outdoor applications, including conduits and plenum installations. Both wires can be for data use.
2	Two #22-AWG twisted-pair solid copper wires in a braided copper tinned shield and 4 additional pairs of #22-AWG solid copper pairs between shield and outer plastic jacket.	Available as that above. The four extra wires are usually used for telephone service.
5	Two optical fibers in a plastic covering jacket.	Available for indoor, outdoor, conduit, and/or plenum applications.
6	Similar to type 1, but wires are #26-AWG wire.	Patch cable or short-run applications.

TABLE 3.4 DEC Connect-Type LAN Cables

Type-Name	Description	Recommended uses
Thin wire ethernet	Small diameter (¼") flexible coaxial cable.	Used to carry digital data at 10 Mbps for short distances. Uses BNC type connectors.
Standard ethernet	Larger size (½") semirigid coaxial cable.	Used to interconnect the above cable to longer distances. Uses BNC type connectors.
Telephone Cable	Standard telephone industry four-wire, twisted, solid-copper, unshielded-plastic-covered telephone wire.	Used for voice/telephone service. Uses modular connector.
Video Cable	Standard closed-circuit television coaxial cable, 75-ohm impedance.	Used for video transmission applications. Uses "F" type connectors.

The telephone industry, along with the Electronic Industry Association (EIA), realized that data could be effectively transmitted over telephone systems at low speed and so defined the RS 232-C standard. Connected to a telco line where any dc voltage and ring power is isolated, digital data can be transmitted. A binary 1 (bit) corresponds to a negative voltage, and binary 0 (bit) corresponds to a positive voltage level. This standard is very common in the United States. The European standard is administered by the CCITT (Telephone-Telegraph International Consultative Committee) and is referred to as Recommendation V.24. The voltage range for a binary 1 is –3 to –25 volts and for binary 0, +3 to +25 volts. An updated, similar specification is referred to as EIA PS-422 and EIA RS 423 for balanced and unbalanced lines.

Both of these standards require a modem between the telephone system or the internal LAN wiring to the computer equipment. The modem has a standard 25-pin D-type connector to the computer equipment data terminal. For best results, this connecting cable should not become too long (less than 25 feet). Table 3.5 gives the pin assignments for the modem to the computer. The computer is referred to as the data terminal equipment (DTE) and the modem is the data communications equipment (DCE). The remarks column shows that there are a lot of control signals doing the asking, accepting, and handshaking necessary to transfer data between equipment.

As mentioned earlier, present data telephone system modems can transmit and receive data at 28.8 Kbps. These modems contain circuits that clean up the data stream and perform a large amount of error checking and so are able to function on relatively poor or noisy telephone lines. The code used by data transferred by a modem is often the American Standard for Information Interchange (ASCII). This well-known seven-bit code is still used today in many business applications in the United States. Figure 3.21 illustrates the form of this code type. Another similar code type is the CCITT number-5 code, which is also a 7-bit code. This code type is more international, and many of the characters are the same as for the ASCII code. The modulation method used by most modems is frequency shift keying (FSK), which works quite well over standard telephone channels. Several solid-state device manufacturers make chip sets for telephone-type modem use. Also, one chip modem device is available. The specification designation is CCITT V.23 or Bell 202. The columns numbered 0 and 1 of the ASCII code are control characters and are listed in Figure 3.22.

3.3.1.2 LAN networks and controls. From the simple computer-to-peripheral equipment interconnect using ASCII code, more standardization was required in order to maintain compatibility to the outside world. When more workstation personal computers were required to connect the many offices to shared printing equipment and mass memory storage, a local area network (LAN) was an obvious solution. The LAN network architecture choices were a bus type, a ring or loop type, or a star type of network topology. These three types of network topologies are shown in Figure 3.23.

The star-type network requires all workstations to communicate to each other through the central controller. The central controller acts as a switching device similar to a telephone system. If workstation 1 is communicating

to workstation 2, workstation 3 cannot interrupt and communicate with either 1 or 2. However, workstation 3 can communicate to any other available
workstation. A bus system allows messages to be transmitted on a common
or bus cable. Since the message should contain the address of the receiving
workstation, only the properly addressed station can take the message. In a
ring-type LAN, message traffic is circulated around the loop. At the arrival
at the nearest workstation, the signal is accepted or retransmitted along the
loop. Each node or workstation acts as a message repeater.

TABLE 3.5 RS-232C 25-Pin D-Type Connector for Modem-to-Computer Terminal

Pin no.	EIA circuit code	CCITT circuit code	Signal type	Remarks
1	AA	101	ground	chassis
2	BA	103	data sent	transmitted data from computer
3	BB	104	data received	modem receives data
4	CA	105	control	computer requests to send data
5	CB	106	control	modem clear to accept data
6	CC	107	control	modem is ready
7	AB	102	ground	signal ground
8	CF	109	control	received line signal (carrier defect)
9	—	—	—	POSITIVE TEST LINE (DO NOT USE)
10	—	—	—	NEGATIVE TEST LINE (DO NOT USE)
11	—	—	—	NOT USED
12	SCF	—	control	secondary carrier defect
13	SCB	—	control	secondary clear to send
14	SBA	—	data	secondary data transmit
15	DB	114	timing	*modem transmit clock
16	SBB	—	data	secondary receive data
17	DD	115	timing	*clock signal received
18	—	—	—	not used
19	SCA	—	control	secondary request to send data
20	CD	108	control	data terminal ready
21	CG	—	control	signal poor (error)
22	CE	125	control	ring/calling indicator
23	CH/CI	—	control	data rate selector
24	DA	—	timing	transmit clock
25	—	—	—	not used

*Used for synchronous modems.

b$_7$			0	0	0	0	1	1	1	1
b$_6$			0	0	1	1	0	0	1	1
b$_5$			0	1	0	1	0	1	0	1
BIT NUMBER						COLUMN				
b$_4$ b$_3$ b$_2$ b$_1$	ROW	0	1	2	3	4	5	6	7
0 0 0 0	0	NUL	DLE	SP	0	@	P	'	p
0 0 0 1	1	SOH	DC1	!	1	A	Q	a	q
0 0 1 0	2	STX	DC2	"	2	B	R	b	r
0 0 1 1	3	ETX	DC3	#	3	C	S	c	s
0 1 0 0	4	EOT	DC4	$	4	D	T	d	t
0 1 0 1	5	ENQ	NAK	%	5	E	U	e	u
0 1 1 0	6	ACK	SYN	&	6	F	V	f	v
0 1 1 1	7	BEL	ETB	'	7	G	W	g	w
1 0 0 0	8	BS	CAN	(	8	H	X	h	x
1 0 0 1	9	HT	EM	)	9	I	y	i	y
1 0 1 0	10	LF	SS	*	:	J	Z	j	z
1 0 1 1	11	VT	ESC	+	;	K	[	k	{
1 1 0 0	12	FF	FS	,	<	L	\	l		
1 1 0 1	13	CR	GS	-	=	M	]	m	}
1 1 1 0	14	SO	RS	.	>	N	^	n	~
1 1 1 1	15	SI	US	/	?	O	_	o	DEL

SP—SPACE
DEL—DELETE

Figure 3.21 ASCII code bit/character chart.

All LAN systems require some sort of access control, which can be either central, distributed, or randomly controlled. Central control uses one designated workstation that controls the entire network access. All workstations have to request access and receive permission to communicate from the controlling computer terminal. Distributed control allows only one workstation terminal at a time to transmit, and this right is passed in turn to all the other stations. Random control requires workstations to listen to the network for a transmitted message, and if none is heard the station is free to transmit.

Control Character	Description	Control Character	Description
NUL	All zeros	DLE	Data link escape
SOH	Start of heading	DC1	Device selector
STX	Start of text	DC2	Device selector
ETX	End of text	DC3	Device selector
EOT	End of transmission	DC4	Device selector
ENQ	Enquiry	NAK	Negative acknowledge
ACK	Acknowledge	SYN	Synchronous idle
BEL	Bell ring (Attention)	ETB	End of transmission block
BS	Back space	CAN	Cancel
HT	Horizontal tabulation	EM	End of medium
LF	Line feed	SS	Start of special sequence
VT	Vertical tabulation	ESC	Escape
FF	Form feed	FS	File separator
CR	Carriage return	GS	Group separator
SO	Shift out	RS	Record separator
SI	Shift in	US	Unit separator

Figure 3.22 ASCII control characters.

Each of these three methods of access control require different methods to implement the proper procedure. Centralized control requires some form of polling of the workstations to perform the network switching and connect the communicating terminals. This type of network uses what is known as the time division multiple access (TDMA) method of control, which means each workstation takes turns transmitting one at a time. Distributed control is often referred to as a token-passing method, which allows the station possessing the token to transmit. This token-passing technique can operate for either ring or bus-type architecture and is referred to as a token ring or token bus method.

All workstations have to sense the token in the form of a carrier sense multiple access with collision avoidance (CSMA/CA) method or a digital token-type message inserted on the ring or bus. Systems using the CSMA/CA require workstation terminals to listen for the carrier signal that is transmitting data, and when the transmission is finished, each station must wait for a specified amount of time. At the end of this waiting time, if no other station starts transmitting, the station can transmit on the bus. Where a token-message method of control is used, the token is a short message that constantly circulates around the ring. When the token is accepted by a workstation, it is not allowed to circulate further. The workstation possessing the token message sends out a token busy message, which now circulates through the ring along with its message. At the finish of transmission, the station reinserts the token, marked free on the ring. The receiving stations' network interface unit sees the busy token, recognizes its own address, reads in the message, and retransmits the whole message on the loop to be received by the sending station, which then transmits the free token. This type of token-ring topology is used by the IBM Token Ring network.

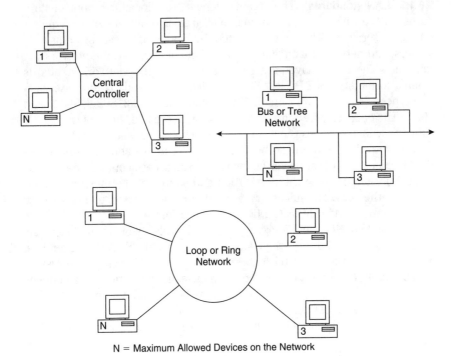

N = Maximum Allowed Devices on the Network

Figure 3.23 LAN network topologies.

Random network control works on either ring or bus technology but has some differences from other methods. This type of network control uses carrier sense multiple access collision detection (CSMA/CD) and is used in the well-known Ethernet LAN system. As mentioned before, sensing a carrier means the workstations are in the listen mode before transmitting. When a work terminal transmits, all other stations receive the message and examine the address, and the addressed station accepts and receives the message. If more than one station starts transmitting, the message format becomes garbled. This is known as a collision. The listening stations detect that a collision has occurred and ignore the message. When the sending stations realize that a collision has occurred, they stop transmitting. To avoid further collision, the transmitting stations have to wait for different periods of time before trying again. This method allows messages to be rapidly transferred if traffic is light.

The previously discussed methods of controlling the networks are acting as the so-called traffic signals controlling the information flow among workstations. LAN systems will be required to interconnect through long distances via the information highway, and commercial businesses will be the largest paying users to the carriers operating the information highway.

3.3.1.3 LAN standards. The organizations setting local area control standards started with the computer manufacturers through customer requests for such services. The software used by many database systems, file servers, and network management controllers determined the network parameters. The IEEE played a major part in developing network standards when it formed the 802 Project Standard. This 802 standard addresses several types of LAN network architectures known as 802.2, 802.3, 802.4, and 802.5. The 802.2 architecture is concerned with the logic link control (LLC) sublayer, which services the layer above it. The LLC sublayer describes the service with the operating software in the layer above and the services on the layer below. This sublayer below is a media access control (MAC) sublayer. Information as to the destination and source of the message is contained in the LLC. The necessary requests and acknowledgments for data are also contained in this 802.2 IEEE specification.

The 802.3 specification addresses the MAC sublayer operating below the LLC layer, which addresses the transmission media of CSMA/CD, token-bus, and token-ring networks. Ethernet operates in this LAN architecture. Data is encapsulated (packetized) for the message-sending station and contains header and trailer information placed at the beginning and end of the data capsule. The tasks contained in the header/trailer contains control information such as synchronization with the receiving station, start/end of data frame, addresses of sending and receiving stations, and error detection and/or correction information. When collisions occur, they have to be detected first, followed by correction instructions to the stations on the network.

The 802.4 addresses token-bus network topology where stations on the bus pass a token from station to station. The General Motors Corporation's MAP architecture conforms to this protocol. This is a broadband RF carrier technique operating in a similar fashion as a cable television system. The RF carriers carrying digital data are transmitted as a midsplit or high-split single-cable coaxial cable bus. The LLC layers contain control information such as passing the token from station to station, maintenance messages for initializing, power-up information, fault detection and correction, and sending and receiving information from the MAC layer.

The 802.4 protocol supports several transmission methods. A single carrier with frequency shift keying (FSK) modulation operates as a baseband signal on a 75-ohm (CATV type) coaxial cable. Drop cable to stations is either 35-ohm or 50-ohm coaxial cable. Data rates of 1 Mbps are transmitted on carriers operating in the frequency range of 3.75 MHz to 6.25 MHz. Another type of LAN system operating at 802.4 is a single-channel (carrier) phase coherent FSK method. The frequency is not continuously moved from one value to another, as in the previously described method, but is essentially switched between two discrete frequencies corresponding to 0 and 1 digital values. The third method is the GM Manufacturing Automation Protocol (MAP), as previously mentioned. Coaxial cable operating as a single cable midsplit or high-split or a dual cable with one cable for each direction are the choices for this method. Data rates of 1, 5, and 10 Mbps are supported by this method.

The 802.5 specification addresses the token-ring architecture, which was developed as the IBM method for an LAN. Transmission is determined by which station gets the token that is passed in sequence around the ring. Data transmission also flows around the ring from station to station in sequence, where each station acts as a repeater. The time of transmission is fixed, and when the time is up, the transmitting station passes the token to the next station. The token is considered free until a station desiring to transmit data receives it. The data transmitted changes the token to busy as long as it is sending data. When the transmitting station is finished with its transmission or its time is expired, it then transmits the free token to the next station. The transmitted data contains address information, packeting information, and error detection and correction direction. The receiving station recognizes its address, checks the packet size and sequence, and checks for errors from the control information sent with the message.

The IEEE 802 specification addressed the early cable-connected LAN architectures and left many areas not considered. LAN users regarded these areas as options. One offshoot of the 802.4 specification was a development of the Manufacturing Automation Protocol (MAP) by GM and Technical and Office Protocols (TOP) by the Boeing Company, which are combined into what is known as the MAP/TOP specification. A reference model for computer network operators called Open Systems Interconnect (OSI) was recommended by the International Standards Organization (ISO) and describes a seven-layer protocol. Both MAP and TOP use the OSI recommendation. These seven layers are the application layer, presentation layer, session layer, transport layer, network layer, data link layer, and the physical layer. Each layer has its assigned duty for initializing, arranging, synchronizing, quality control, routing, node-to-node routing, and data-bit-stream transmission on the cable.

Data transmission using a star network topology was previously introduced. The network control function was operated by a controller that acted like a private branch exchange (PPX). This type of PBX network used telephone-type cabling, and in many cases, offices operating with LAN were also wired with PBX cable. The PBX cable carried the normal telephone traffic and computer data through a modem. The workstation could also operate on the LAN for interoffice computer traffic. This type of operation led to bridging or connecting various LANs together through telephone-type modems through the local telephone company's PBX system.

3.3.2 WAN/MAN wide and metropolitan area networks

LAN systems connected into a larger system became known as a wide area networks (WANs). Some large LAN systems operating on a campus or industrial park also became known as WANs. An often-used method of connecting LAN systems to form a WAN is the use of a direct-dial system or a leased telephone line. Data rates for the dial-up method are presently 28.8 Kbps, and for leased lines, 56 to 100 Kbps, depending on the quality of the line. At present, with improved equipment and line quality, data rates of 2 Mbps can be realized. WAN technology is used to connect LANs belonging to a company or corporation. Where a public network is concerned, the interconnection is known as a metropolitan area network (MAN) and usually contains the transmission of voice, data, and video signals. The technology for WAN and MAN interconnection is often the same. However, WANs are private and usually commercial, while MANs involve public use.

3.3.2.1 Network topologies.
As we have seen, the network topology of many LAN systems differ. Therefore the interconnection needed to transfer data among differing LAN types requires the use of special interface equipment. This network interface equipment essentially has to intercept messages from the sending LAN and convert them to a format for the receiving LAN. The control portions of the message containing source, destination, packeting, synchronizing, parity, and error-detection information all have to be changed to the receiving LAN format. For a star-switch type of network, the interconnection of the controller is all that is required. For bus or ring-type topologies, the token-passing information portion of the message will most likely have to be changed because the tokens will be different. MAN systems often use the public switched telephone network (PSTN) for the interconnect. Therefore the use of telephone modems simplify the intercon-

nection. Where leased lines are used, the interconnection will require a custom-designed network interface unit between the line and data terminal.

3.3.2.2 Standards and interconnections. There are standards for defining the interconnections forming WAN/MAN systems. Where a dedicated, leased telephone line is used, the standard is known as X.25, which is a data packet switching method. This standard is recommended by the International Telegraph Telephone Consultative Committee (CCITT) for data terminal equipment (DTE) and data communication terminal (DCE) interconnection using packetized data. This committee is a part of the International Telecommunications Union (ITU), a member of the United Nations. Much of the development work leading to this standard was done by AT&T in the United States and was accepted by the CCITT.

Widespread use of optical fiber by the telephone industry made possible the use of leased fiber lines for the WAN/MAN interconnect. This allowed data transfer rates of transmission to skyrocket, operating at about 150 Mbps for distance of up to 100 kilometers.

Some WAN/MAN systems operate in a similar fashion as a cable television system using single or dual cable methods. A standard for this type of WAN/MAN has been worked out by the IEEE and is known as the 802.6 standard. This standard allows a broadband system to carry data, voice, and video signals on RF-modulated carriers transmitted on a coaxial cable. This 802.6 standard was developed from the MAP-type system previously described and operates using a dual bus-type network topology known as a distributed queue dual bus (DQDB) system.

3.3.2.3 Operations and maintenance considerations. Operation and maintenance of WAN/MAN systems is far more difficult than for LAN systems, simply because the distances are longer, the equipment is more diverse, and it often involves several interconnected systems. For broadband MAP-type systems, carriers appear when data is transmitted and disappear when transmission ceases. Therefore, testing of carrier level is next to impossible because now you see it, now you don't. Various test instruments have had to be developed in response to the needs of such network maintenance problems. Where the telephone system is used as an interconnect method, the telco maintenance crew will handle maintenance and repair problems. Private maintenance contractors also are available to perform maintenance procedures for the other portions of the interconnect facilities. Instruments that check for end-to-end data quality are known as bit error rate testers (BERT), and they are extremely useful. These devices transmit a known pattern of digital bits on the system and then receive them on the return line for error testing. This overall data-quality check is a good measure of the system's communications quality. Data transfer quality is often specified as to the bit error rate, making this type of instrument's use very attractive as a signal-monitoring device.

Cable System Signals and Interconnections

In the past, the electrical signals making up messages have been many and varied. When the signals are determined as continuous, time-amplitude varying signals, they are defined as analog. When the electrical signals are pulsed on or off, corresponding to binary ones and zeros, they are defined as digital. These signals make up the so-called baseband signals containing the message information. Baseband signals can be transmitted directly along conducting wires. However, as the frequency or time rate of change increases, the cable loss increases. Therefore when more message traffic is being squeezed into a shorter period of time, the cable loss increases, thus attenuating the signal. To increase the data throughput to a cable system, modulation techniques and repeater amplifiers are employed. This chapter investigates the signaling methods used to transfer voice, video, and computer information from point to point. This is where the information highway is at present.

4.1 Telephone Signals/Message Traffic

The public switched telephone network (PSTN) in the United States was developed through the invention of the telephone by Alexander Graham Bell in 1820. From this rudimentary device came the incredible development of the American Telephone and Telegraph Company (AT&T) known as the Bell System. The guidelines for the company were service and reliability. Equipment was to be designed for a 40-year useful life, and system reliability was a rule. Such difficult design and manufacturing specifications became too much of a problem for many companies, so the telephone company started its own manufacturing company for telephone equipment—the Western Electric Company. The research branch of AT&T was known as Bell Labs. Research carried on by Bell Labs made many important inventions and contributions to the industry and general public.

The main product of the telephone company was voice communications throughout the country. Research into hearing and speech perceptibility provided constant improvements in voice communication. Today, telephone service is a large factor in the way we do business and carry on our social life. With the development of computer equipment and ultimately the personal computer (PC), the telephone industry responded with methods and equipment to transfer digital signals on the telephone lines. The device known as a telephone data modem was developed. The telephone development progressed to overseas with systems being built throughout most of Europe.

Because the Bell System developed in the United States was the first phone system, European development adopted much of the American methods and standards. System compatibility was never a problem until the first undersea cable for telephone use was installed. Previously, undersea telegraph cable was the communication medium between America and Europe. Morse code was used for both manual and mechanical methods for receiving and sending the signals. Since the European standards were derived from the American Standards, the compatibility issue was minimized.

Development of the telephone plant in the United States involved many areas. Cable unaffected by moisture, the dial telephone, long-haul trunking lines, and new techniques enabled the telephone system to expand coast to coast. Reliability has always been an important subject with the Bell System. Therefore, technical training for maintenance and repair personnel has been carried on by the company as new developments in instruments, tools, and procedures have been made. It was not so very long ago that the local telephone exchanges were strictly electromechanical in nature, and the first coast-to-coast direct-dial call was placed in 1951. Present-day telephone systems have electronic telephone sets, microprocessor solid-state exchange switching systems, satellite and fiber-optic long-distance trunking, plus a host of new test instruments. The telephone network covering North America is developing a system capable of transmitting voice, video, and high-speed computer digital data to businesses and homes requiring such services.

4.1.1 Audio/Voice Signals

The "meat and potatoes" of the telephone industry is basic voice traffic. Tests at Bell Labs long ago found that the greatest part of voice intelligibility is confined in the frequency band of 300 to 3000 Hz. Thus the band of 4000 Hz, which includes this band plus some extra space for control signals, is defined as a single-telephone one-way channel. The telephone set deletes frequencies above 3000 Hz. Therefore the control tones are not heard by listeners and do not interfere with telephone conversations.

The twisted-pair telephone cables operate well at low audio frequencies, which eliminates any need for amplification at the local exchange level. Trunking and long-haul applications use a mix of coaxial cable carrier systems, multichannel microwave radio links, and fiber-optic cable systems, which use amplification in their processes. The conversion of the low-frequency (300–3000-Hz) voice band, analog-to-digital signals was studied many years ago at Bell Labs. Much research into the theory of communications by Bell Labs provided a large amount of information that proved to be beneficial when the need for data communications arrived.

Essentially, voice channels were digitized by solid-state analog-to-digital converters, transmitted as digital data, and at the receiving point converted back to the analog state to operate the basic telephone set. This, of course, is an oversimplification of a complicated process. Several one-way telephone conversations are digitized and multiplexed together and transmitted. At the decoding or receiving station, the messages are separated and routed on their way. Each voice channel has a word position in a frame. Thus the decoder examines and assembles the words contained in a voice channel. This procedure is known as pulse code modulation (PCM) and is used on coaxial-cable long-haul cable systems, microwave radio links, and fiber-optic systems.

4.1.1.1 The telephone set. The one main piece of customer equipment is the basic telephone set or instrument. This device was and is designed for a 40-year useful life. However, since the so-called Bell breakup, many varieties of telephones have appeared, and as most of us know, some are really of very poor quality. Present-day telephone equipment is required to operate in the pulse dial and touch-tone mode because in some rural areas the local exchanges still operate with early electromechanical switch banks.

The rotary dial is an early form of a serial pulse generator that placed current interruptions on the line corresponding to the number dialed. These pulses caused the electromechanical relays to search for the dialed number line in the local exchange. A diagram for a conventional telephone set with pulse dial is shown in Figure 4.1. When the handset is removed from its cradle, the switch hook contacts close, causing a loop current to flow from the 48-volt battery B through line relay RLY and dropping resistor R. The activation of RLY in the local exchange station connects the telephone instrument to an open line, which sends back a dial tone to the telephone set. The dialed number causes the local exchange switch bank to select the line of the called party. When the connection is made and no loop current is de-

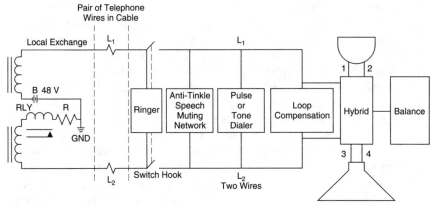

Figure 4.1 Basic telephone set.

tected on the called party's line, the ringing circuit is activated. The ringing tone is also sent back to the calling party. When the called party answers the call by removing the handset, the resulting loop current holds the connection while the telephone conversation continues.

If the called party is conducting another telephone call when another call comes through, the loop current presence activates a busy signal to calling parties. Early touch-tone telephone sets had electromechanical tone rods or bars used to produce the dual tones. The hybrid device in the electromechanical telephone set was a matrix of one-to-one transformers that essentially split the two telephone wires to a four-wire system used to connect the mouth and earpieces.

This circuit is shown in Figure 4.2. The winding sense of the four transformers allows for simultaneous talking and listening on one pair of wires, L_1 and L_2. When a user is speaking into the mouthpiece, current from ter-

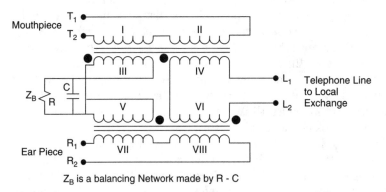

Figure 4.2 Transformer type telephone hybrid circuit.

minals T_1 and T_2 flows through windings I and II, causing a voltage to be induced in windings III and IV. This voltage causes current to flow in lines L_1 and L_2 through windings III, IV, V, and VI. The winding sense of III and VII and IV and VI are opposite, causing cancellation of voltage and disallowing voice currents to reach the earpiece.

For the reverse condition, when the called telephone causes currents to flow in L_1 and L_2, the hybrid does not allow the current to reach the mouthpiece—only the earpiece. By slightly unbalancing the hybrid, 100 percent cancellation is not achieved, which allows some talking currents to reach the earpiece. This is referred to as a *side tone* and allows the person speaking to hear his or her voice in the earpiece. This prevents a person from shouting into the mouthpiece and helps level the voice signals on the lines.

The antitinkle circuit is actually a form of surge protector that prevents the bell from clinking when the dial is being operated. The loop compensation circuit acts as an impedance-matching network and voice frequency filter. This type of telephone set was physically large and quite heavy due to the hybrid circuit. The main attribute of this telephone was its ruggedness and high reliability. Even the cords connecting the handset were extremely reliable, mainly due to their highly flexible construction.

The electronic telephone set was developed in recent years and many versions are in use. All the functions of the basic telephone set are accomplished by electronic devices in the present-day electronic telephone. The action of the hybrid is synthesized on a solid-state microchip along with speech circuits and tone/pulse dialing. A circuit of one type of electronic telephone is shown in Figure 4.3. Since electronic solid-state devices operate at low-value, direct-current voltages, protective devices are built into the electronic telephone to control voltage surges and prevent damage due to incorrect connections to the lines.

Present-day electronic telephone sets are smaller and lighter due to the miniaturization offered by solid-state chip devices. These telephones run the spectrum from cheap, poorly manufactured devices to reliable, rugged telephones like the older electromechanical telephones. The more high-quality electronic telephones provide more features, a pleasant ringing, amplification for the hearing impaired, stored-number memories, and redial last number, to name a few. Business telephone systems found in many present-day offices are highly sophisticated, electronically controlled switching systems capable of multiline inputs to multiline outputs for various office locations. Electronically generated voices and solid-state memory devices are found in present-day catalog-shopping telephone systems. Local telephone exchanges contain microprocessor-controlled routing switches with telephone-number memory banks that are completely automatic, thus the phrase "no operator involved." Timing of call duration and line-rate information feeds the computer-controlled accounting system, which determines the rates, charges and tolls for monthly billing.

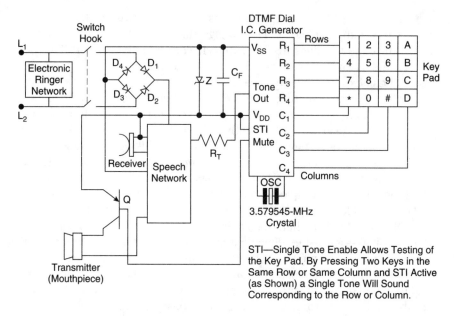

Figure 4.3 Electronic telephone set.

4.1.1.2 Computer modems.

Since it was determined that the vast telephone system in North America could also provide data communications by use of an interface device called a telephone modem, the telephone industry has and is in the computer data transmission business. By today's standards, early modems operated quite slowly—at rates of 300–1200 bps for full-duplex operation.

The heart of modem circuitry is the microchip called a universal asynchronous receiver-transmitter or UART. Essentially, this microchip receives the parallel output data from the computer terminal and converts it to a standard serial bit stream to feed the telephone-line driving section connected to the telephone line. This same device receives serial data from the telephone line and feeds it through the UART for conversion to parallel data required by the computer terminal.

The word modem is a contraction of the words modulator-demodulator. Modems use either the 7-bit ASCII code or the 5-bit CCITT code for transferring alphanumeric digital data. Telephone modems use a carrier that is modulated with the digital data bits. The usual method is called frequency shift keying (FSK), where, for example a low-frequency value represents a binary zero and a high-frequency value represents a binary one. This is often referred to as two-tone modulation by telephone people since a single frequency is regarded as a tone.

Phase modulation is also a technique used in modems. One phase condition represents a zero and a second phase a one. Early modems were connected to the telephone system by a cradle to hold the telephone handset. Audio tones shifted in frequency were acoustically coupled through the handset by the cradle device driven by the modem. The transmit/receive section of a modem uses a microchip that receives and sends serial data (a bit stream) to the UART for conversion to parallel data.

A microchip interface is shown in Figure 4.4. Notice that normal telephone functions such as ring detection on/off-hook condition, and the hybrid functions are handled by this section of the modem. This section uses the serial bit stream provided by the UART and shifts a frequency derived

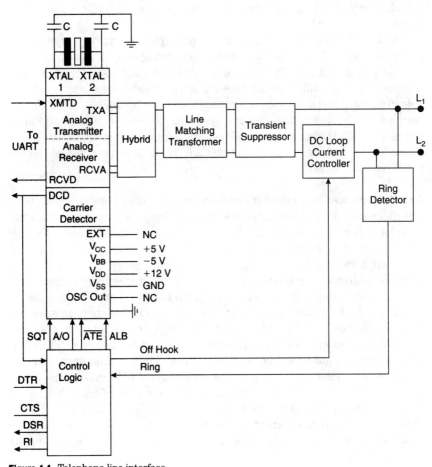

Figure 4.4 Telephone line interface.

from the crystal (XTAL) oscillator between two different values to perform the FSK modulation. The FSK-modulated data received from the line through the hybrid is demodulated in the inverse manner by the microchip interface and provides a serial bit stream to the UART.

Present-day modems operate normally at full-duplex rates of 14.4 Kbps, and the latest varieties operate at 28.8 Kbps. This might seem to push the telephone system to the limit. However, the newest telephone modems contain signal-processing circuitry that essentially cleans up the data signal by noise reduction and signal reconstruction techniques and with error detection and correction circuitry. Reliable higher data rates are achieved. Conditioned leased lines can also be used at even higher data rates, depending on the line quality. Such lines might not contain the 48-volt power source or ring power normally appearing on telephone lines. These lines are referred to as *dry lines*. Normal LAN/WAN bidirectional interface equipment can be used to achieve higher-speed data rates.

Normal telephone modems running in full-duplex operation carrying computer data and regular voice telephone service might seem like an information highway, and it actually is one. It is only lacking video services. However, there are a few manufacturers of video slow-scan telephone modems that are able to transmit still or slow-moving images. The area of telephone system operations that limits the transmission speed requirements of video and high-speed data is the twisted-pair sections of the telephone system, which appear at the local exchange level. Indications are that this will be changed in the not-too-distant future.

4.1.1.3 Dedicated audio lines (radio service).

Dedicated dry-type telephone leased lines or circuits have been and still are being used by many radio stations for either the studio-to-transmitter feeds, a status monitoring telemeter system, or both. The transmission facilities involve the high-power transmitter electronics, the transmitting tower, and emergency power-generating equipment often placed on the outer perimeter of urban-suburban areas. The studio, on the other hand, is often located in the heart of cities and towns. The status of the remote location with regard to security conditions, equipment operation, and fire alarm are important to the station operator. Measurement information for transmitter power and frequency, antenna current, and tower light conditions can be monitored by a telemeter modem driving a dedicated line back to the studio, where personnel can be advised. The program audio is fed to the transmitting location on a pair of dedicated leased lines. Such lines can supply signals up to about 10 kHz, which is ample for AM radio operations. Some dedicated lines can operate at even higher frequencies, and the telephone industry is well known for its extensive line testing and evaluation program.

4.1.1.4 Digitized voice signals.
As earlier stated, voice telephone transmission is the meat and potatoes of the telephone industry. The Bell System has researched very well the digital conversion of voice audio signals. Analog-to-digital (ADC) and digital-to-analog (DAC) converters have been well known to the telephone industry for many years. Present-day solid-state technology has produced chip sets that can perform the ADC and DAC functions to amazingly high speeds. The conversion of voltage levels into a digital-data word is the job of the ADC. If a voltage level is described by one bit, only two levels can be represented. With the addition of one more bit (two bits), four levels can be represented, and with three bits, eight levels can be represented. Thus, a 2^n situation is realized.

If a 0–1 volt level is converted to an 8-bit digital number, and since 2^8 corresponds to 256 levels, each level or $\frac{1}{256}$ of a volt corresponds to 0.00391 volts or 3.91 mV per bit.

This is the voltage resolution and is less than $\frac{1}{2}$%. This is illustrated in Figure 4.5.

For voice work, eight bits is ample. The next consideration is the speed of conversion as related to the frequency of the signal to be digitized. This has to do with the speed of sampling the analog audio voltage waveform. Examine Figure 4.5, sample number 1. If represented by one bit, this sample will have a digital value of 0, for two bits (01), and (011) for three bits. The audio signal varies considerably between sample 1 and 2, 2 and 3, etc. What is needed is an increase in the number of samples per unit of time. If

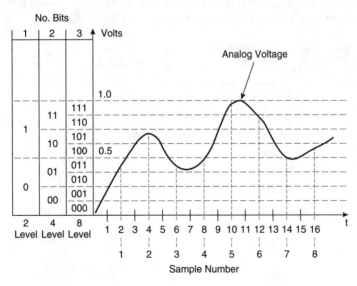

Figure 4.5 Voltage resolution and number of bits.

the number of samples for the same time is doubled, i.e., 16 samples in the same space, then smaller voltage variations occur from sample to sample.

The main part of the conversion process from analog to digital is the sample-and-hold circuit, which takes a slice of the voltage level and holds it long enough to be converted into a digital value. Then another slice is made, and conversion is made in sequence. During the holding time, the analog value might change. If the sampling rate is fast enough, the lower-frequency voltage variations will be insignificant. A rule of thumb is if the sample and hold time are equal, the sampling frequency should be at a minimum of two times the highest frequency to be digitized. For the upper voice limit of 3 kHz, a sampling speed of 6 kHz will suffice. An 8-bit digital value will occur at a 15-kHz rate for one-half the period.

Some present-day flash-type analog-to-digital converters do not require a sample-and-hold circuit, and they convert as soon as sampled. Very high bit rates (100 Mbps) are possible, but a maximum 10-bit resolution limits some of its usefulness in certain applications. For audio voice circuits, the lower sampling rates and 8-bit resolution is sufficient. For voice telephone systems, the first bit is the most significant bit (MSB) and contains voltage polarity information. The remaining 7 bits correspond to the amplitude of the voltage sample, where there are 128 quantization voltage steps. Appendix J further investigates the voice frequency digitization process. The formula that describes the signal-to-noise ratio for a given number of binary bits provides the information contained in Table 4.1. If more bits per time slot are required, the sampling rate and hence bit rate has to increase. This requires more frequency bandwidth from the transmission link. However, for telephone applications, a higher sampling frequency of 8 kHz is chosen.

TABLE 4.1 **Number of Quantizing Steps and Signal-to-Noise Ratios, Dependency on Number of Bits in a Code Word**

n bits in the code	Number of quantizing steps	s/n dB
7	128	43.76
8	256	49.76
9	512	55.76
10	1024	61.76
11	2048	67.76
12	4096	73.76
13	8192	79.76
14	16384	85.76
15	32768	91.76
16	65536	97.76

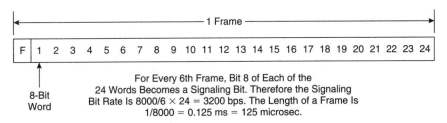

Figure 4.6 T-1, DSI 24-channel digital frame.

Voice channels containing telephone speech signals are combined into a digital frame of bits. This frame has been standardized to some degree by the telephone industry and contains 24 eight-bit words corresponding to 24 one-way (half duplex) telephone service. The other half of the telephone conversation traveling in the reverse direction has to have its place in a 24-channel system. If 24 channels of 8-bit code words occur at a rate of 8000 per second (8 kHz), then 24 words × 8 bits/word = 192 bits. Also, if one bit is allowed for a frame-marking bit, then the result is 193 bits in a given frame. This occurs at a rate of 8000 times a second, which results in a bit rate of 193 bits × 8000/seconds = 1,544,000 bits/second or 1.544 Mbps. This is designated as T-1 standard, which later was referred to as Digital Standard 1 (DS1) by the telephone industry and is illustrated in Figure 4.6.

The bit stream of the 193 bits can be represented by a pulsed electrical signal called a line code. These signals can represent the bits as a pulse train of nonreturn to zero (NRZ) or return to zero RZ. For the NRZ case, the pulses represent a binary 1 not returning to the zero voltage state until the end of the synchronizing clock signal. In the RZ case, the pulse for a 1 bit does return to zero during a clock pulse. This is illustrated in Figure 4.7. Also illustrated in the figure is a bipolar-type code referred to as alternate mark inversion (AMI), where alternate binary ones reverse voltage polarity. The telephone industry in the United States developed from the beginning T-1 standard, a series of standards for stacking various numbers of DS-1 levels to form a hierarchy of digital multiplexing. This standard is asynchronous in that the sampling rates of the stacked levels are not synchronized. Table 4.2 summarizes these levels and the developed hierarchy.

The method required to transmit the digital-signal bit stream varies as determined by the data bit stream. Radio methods are performed by microwave radio links and coaxial cable, which uses RF carriers modulated by the bit stream. Fiber-optic methods employ optical transmitters and receivers modulated with the digital data. Often voice channels in the DS hierarchy use various transmission types on their way to the end user. The system routing the half of the full-duplex telephone conversation has to be routed through a similar path going in the reverse direction so the accumulated delays would be

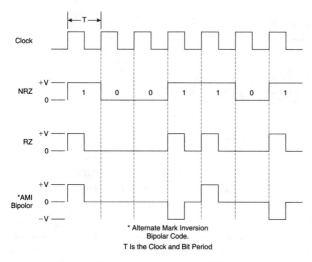

Figure 4.7 Pulse code modulation signal for NRZ, RZ, and bipolar codes.

TABLE 4.2 Digital Levels Used by the Telephone Industry

Level designation	Numbers of voice channels	Bit rate of data stream Mbps	Type of line code	Transmission medium
DS1	24	1.544	Bipolar	T-1 radio/cable
DS1C	48	3.152	Bipolar	T-1c cable
DS2	96	6.132	Bipolar	T-2 cable
DS3	672	44.736	Bipolar	Radio, fiber, optical
DS4	4032	274.176	Bipolar	Coaxial cable, fiber, optical, radio

equal or nearly so. Notice that when the number of channels are doubled between DS1, DS1C, and DS2, the bit rates are less than doubled. This is because the added frame-marker signaling and control bits are not doubled. The layering of frames from one level to the other is a complicated process and is discussed at length in many of the references given in the bibliography. Since the method is asynchronous, to go from DS4 to DS1 the whole process has to be unlayered from DS4 to DS3, then to DS2, DS1C, and DS1.

DS1C is essentially made up from two DS1 layers. The AT&T system uses this digital hierarchy with various cable and radio techniques.

When fiber-optic systems became practical, the telephone industry, which was responsible for much of the development, was more than ready to implement the technique for voice digitized service. The DS1-DS4 digital methods were used to drive the fiber-optic laser transmitters. The bit stream of PCM voice channels was used to control the optical power output of the transmitter, which converted the information to light energy. At the receiver, the light energy was converted back to the binary coded PCM in bit stream, where the layers were separated out and the various voice channels routed to the proper destination between the calling parties.

To construct the DS4 level from the lower levels, beginning with DS1, the process of multiplexing to one party and the demultiplexing from the other calling party was often referred to as a MULDEM.

This was a contraction of the words multiplexer/demultiplexor. For full-duplex service, needed to complete a telephone conversation between called parties, two optical-fibers, one for each direction, was needed. A pair of MULDEMS was required at each end. The block diagram in Figure 4.8 illustrates the concept. This method of voice transmission in full-duplex has been employed by the telephone industry since about the mid to late 1980s. Since telephone voice signals are digitized at various locations, the digital signals are unsynchronized (asynchronous), and each level (DS1, etc.) has its own bit transmission rate. Also, each level carries coded bits, referred to as overhead bits, which carry important control information. The overhead bits contain such information as data frame, layer, and routing identifications. To efficiently use fiber-optic methods to carry the lower-order signal levels, the means of packing those levels into synchronous packages is referred to as a Synchronous Optical Network (SONET). This is a standard that has been developed specifically for optical fiber transmission that enables the packaging of a synchronous level. This means that DS1, DS1C, DS2, and DS3 can be transmitted as a package. Since DS3 is made from DS2, DS1C, and DS1, this level will have to be unlayered to get to a DS1 level. With the SONET technique, since it is synchronous, each DS3 bundle can be handled on and off the fiber separately from other DS3 bundles. In other words, the SONET method does not have to be unlayered. The SONET system is often likened to a commuter train where each person (payload) can pick any car and sit anywhere and get off at any designated station. The basic SONET signal has a data rate of 51.84 Mbps and can be multiplexed up to higher rates by multiples of this basic rate. This SONET

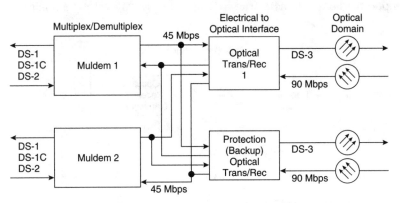

Figure 4.8 Fiber-optic multiplexing-demultiplexing system. Full duplex.

TABLE 4.3 SONET Optical Carrier Designations and Line Rates

Optical carrier rate no.	Data line rate (Mbps)	Remarks
OC1	51.84	Basic building-block OC1
OC3	155.52	3 × OC1
OC9	466.56	9 × OC1
OC12	622.08	12 × OC1
OC18	933.12	18 × OC1
OC24	1244.16	24 × OC1
OC36	1866.24	36 × OC1
OC48	2488.32	48 × OC1

format drives an optical carrier system at various levels, i.e., data bit rates. (See Table 4.3).

These higher-rate levels made from the basic OC1 building block have to contain overhead data on the routing, framing, and layering identification. The SONET layers of digital information fall into four main categories, which are path, line, section, and optical or photonic. Figure 4.9 illustrates the layers as per equipment layout. The path layer deals with the input/output path information, which is the assemblage of DS1s and DS3s containing voice and video digital signals and synchronizing into proper format. Mapping of the DS1s–DS3s also appears in this overhead layer. The line layer contains information on assembling the payload contained in the path layer, along with path layer overhead, and the line layer converts the payload to an optical signal. The section layer contains information on the optical transmission, such as framing and/or signal scrambling methods. The actual photonic layer (optical) transmitter-receiver pairs require no overhead bits.

The whole SONET system has developed into a Synchronous Digital Hierarchy (SDH) and is referred to in the industry as SDH/SONET. The fiber-optic medium acts as a true information highway with the ability to

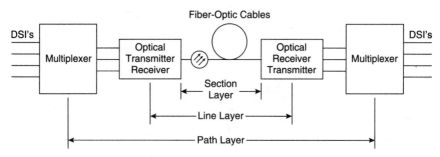

Figure 4.9 SONET overhead layers as related to transmission equipment.

carry multiple voice channels, computer digital data, and digital video in a bidirectional or full-duplex manner. The data addition and drop-off points have to take place at the path-layer terminating equipment, thus acting as an access ramp to the main highway. The DS1 signals can arrive at the path-terminating layer by microwave radio link, coaxial cable system, or optic fiber.

4.1.1.5 Overseas telephone services. The telephone industry has provided undersea telephone service for many years. Of course, early systems were quite crude, did not permit many calls, and were very expensive. However, as better-quality cables and amplifying equipment were developed, the service improved.

Today most telephone service between countries is accomplished using satellite technology. Many telephone calls using the previously discussed DS1–DS3 digital methods were easily integrated into a satellite up-down full-duplex system. This brought the cost of telephone voice service lower, making the service affordable to many more people.

Satellite technology is not without its problems. Semiannual eclipses, often referred to as sun outages, and weather conditions affect signal-to-noise ratios sufficiently to affect overall communication reliability. Since all orbiting satellites do not have sun outages at the same time, satellite channels can be switched from affected satellites to ones not having an eclipse. When fiber-optic cable methods became more fully developed, the telephone industry installed submarine fiber-optic cables. Many countries separated by oceans have installed fiber-optic submarine cable systems. The world's longest fiber-optic submarine cable is 18000 km (11250 miles) and connects Singapore to Marseilles, France. This venture involved many countries, suppliers, and personnel. The cost was approximately 730 million dollars and consisted of seven segments using 137 optical repeaters. The bit rate is at 360 Mbps per pair of fibers, with an error rate of less than 10^{-10} per repeater span, which is 130 km each. The optical wavelength is 1550 nanometers. Each repeater station operates at full redundancy.

4.1.2 Video and miscellaneous services

The telephone companies (mostly the Bell System, which is the largest company) are still the largest carriers of communication services, mostly in the form of plain old telephone service (POTS). When the broadcast television industry came to be developed, the telephone companies were called upon to provide video service. The radio broadcast industry often used leased telephone lines from the in-town studio to the remote transmitter site. Also, during the times of the cold war, when national security was a major concern, the Department of Defense developed remote radar sites to scan the borders of the countries for any hostile action. Radar video information was gathered from remote radar installations and sent via microwave radio to a specified, secure area. The telephone industry was contracted to install such microwave radio links to provide secure surveillance information to the secure location. The transmission of video information in the form of standard television signals and radar scanning signals gave the telephone technical people much experience, which became extremely useful at a later time.

4.1.2.1 Network television video services. The telephone industry was one of the first providers of video service to the network television industry. East-west communications via microwave radio gave the networks coast-to-coast studio communications. Drop/add service along the way allowed the networks to provide the affiliated stations with network program material. Also, news and program information could be added and dropped to the appropriate station affiliates. With experience gained from providing this service, the telephone industry set the preliminary standards for television video/audio transmission. These standards, pertaining to signal quality, were developed using test signals that were later added to the video vertical blanking interval known as *vertical interval test signals* or VITS. These test signals enable technical personnel to identify any problems with the video signal and are used to indicate which piece of equipment might cause the problems. Video transmission using microwave radio techniques was analog in nature, and the video signal modulated a microwave radio carrier. Digital transmission of video pictures has since been developed.

4.1.3 Interconnects to the telephone network

The telephone system in the United States is an extremely complex network of wires and cables of many types that are used for a variety of purposes. The local exchange consists of copper twisted-pair cable. Trunking methods connecting to adjoining area local exchanges often employ hard-wire twisted-pair cables. When distances become greater and connections to higher-level offices are involved, then coaxial cable methods are often the technique of choice. When the distances become larger and there are no connecting pole routes or roads, then microwave radio methods are used. When service involves crossing oceans, submarine cables or satellite techniques become the methods of choice. Since hard conductive submarine cables were few in number not many years ago, satellite communications offered the only solution for many foreign-country communications.

4.1.3.1 Hard-wire trunk lines. Hard-wire trunk lines used to connect the various-level offices were simple and straightforward because the technology was essentially the same as at the local exchange level. The switching systems and the power source for the loop current were essentially the same as for the local exchange system.

Sometimes larger copper wire was used and different equalizers were employed to work the longer distances. Also, amplifying equipment was also available and used when needed. The maintenance of such trunk lines did not present many different problems than for the local exchange system. Identifying which cables in aerial or underground plant were trunk or local lines could be a problem. However, colored tracer markings often identified which cable was which. Presently, most telephone maintenance areas have their maps recorded on CD-ROM and maintenance trucks are often equipped with laptop computers with CD-ROM drives for tracking the plant and troubleshooting.

4.1.3.2 Coaxial cable long lines. As discussed in chapter 3, coaxial lines showed how telephone conversations were analog-transmitted by radio frequency carrier modulation techniques. Connections to local telephone exchanges from these coaxial systems took place at each end of the run or at areas where there was a repeater. At each repeater site or at the sending or receiving end, service to local exchanges was possible. Telephone signals could be added or dropped by demodulation/remodulation of the carrier with the selected audio telephone signals. The signal to be added was used to modulate a vacant carrier, often vacated by a signal being dropped. The switching systems and the line selections were crucial in assembling both parts of the full-duplex telephone conversation. Essentially, the path lengths had to be nearly the same so no signal echo would result. It is simpler to select equal-length paths than to delay the portion of the signal that arrived first.

4.1.3.3 Fiber-optic cable link. Using the SDH/SONET method of stacking and layering, fiber-optic sections of the telephone plant have enormous capacity to carry telephone service. This type of signal transmission is the high-speed bit-stream modulation of the light energy output of lasers. The high-speed bit stream consists of overhead bytes that identify the layer's data payload envelopes and control information. In some areas, DS1 through DS3 also uses microwave radio, where the bit stream of digital data modulates RF carriers. Eventually the digital payload is unloaded from SONET into the DS formats, which are further unlayered into data words corresponding to telephone message traffic.

The digital sections of each part of a telephone conversation have to be assembled and converted to the analog electrical power required by the local telephone exchange. Number called, number of calling party, time of call, and duration of call are contained in the DS overhead information.

Since computer digital information is transmitted by the telephone system by use of the modem interface units, the telephone industry has installed equipment that places modem traffic into the DS format and ultimately on a SONET fiber-optic communication system. Data rates from end-to-end users through the present-day telephone system are up to 28.8 Kbps. Many of us have built-in PC modems with bit rates of 14.4 Kbps and have had them for several years.

4.1.3.4 Microwave interconnects. Microwave radio techniques have been used by the telephone industry for many years. Several modulation techniques have been employed in the past to load telephone traffic on a microwave RF carrier. The carrier method mentioned in chapter 3 for single-sideband suppressed-carrier (SSBSC) modulation was used on coaxial cable and on microwave carriers. Microwave paths operated at one frequency for one direction and another frequency for the other direction. Thus the transmitters-receivers were separated by frequency. Paths crossing each other operated on a different pair of transmit/receive frequencies, thus avoiding interference. The telephone industry maintains proper records of the path locations, intersections, and frequencies required by the FCC. When digital methods were developed for telephone voice transmission, the microwave radio system was easily adapted to digital transmission. The number of telephone channels using the DS formats on microwave radio was greater than the number for the analog SSBSC method, thus improving telephone communication efficiency. Since microwave links were bidirectional, both halves of the telephone conversation were channeled along the same path so signal echoes would not build up. The same goes for digital two-way, full-duplex communications between computer equipment.

Microwave radio path engineering is very important in order to achieve superior communications. Signal-to-noise ratio is extremely important, as is signal distortion. Rain fade and snow accumulation on antenna equipment affect signal-to-noise ratio. Also, electrical interference, both man-made and natural (lightning), can and do cause problems. Since many microwave repeater locations are remote, the loss of electrical power is very important. Therefore, standby power systems in the form of automatic engine plants, either natural gas fired or diesel powered, are the usual choices. These electrical generators can keep the microwave site operational as long as the plant's fuel supply lasts. When commercial power returns, the standby plants shut down. These sites are monitored by technical personnel on a regularly scheduled basis and a log is kept of the maintenance procedures performed.

A microwave path engineering study is given in appendix K. This is an elementary case study to illustrate what the problems are and their causes. Usually when a microwave radio link is under construction, periodic testing will indicate whether proper operation is achieved. As each section or hop is turned on and equipment is operational, a signal of the type the link will transmit is applied and the quality is tested at the receiving end. When digital signals are used, a bit error rate tester (BERT) is often used. This instrument has the capability of sending a bit stream at variable rates, and when the return path returns the signal, the BERT will make a measurement of the bit error rate (BER). BER is defined as:

$$BER = \frac{\text{no of false bits}}{\text{no of received bits}}$$

For a system with a 2.048-Mbps rate, 1 error bit amounts to a BER of:

$$\text{BER} = \frac{1}{2.048 \times 10^6} = 4.9 \times 10^{-7} \quad (\text{approximately 1 bit error in 2 million})$$

If, on the other hand, video television signals are to be used on the link, then video baseband signal quality measurements should be made at both ends of the link.

4.1.3.5 Satellite-overseas connections. The telephone industry was quick to place satellite technology in operation. The efficiency and ease of installing uplink and downlink operations made this a very cost-effective method. This technology for full-duplex operations used one band of frequencies for the uplink (transmitter) at one end and another band of frequencies for the transmitter at the other end. This technique for a simple up/down full-duplex system is shown in Figure 4.10. Each earth station had its own transmitting band and one or more receiving bands. The method of forming groups of half-duplex telephone conversations into master and supergroups discussed in chapter 3 was used on the satellite link, mainly because the technology was there. When digital methods were developed,

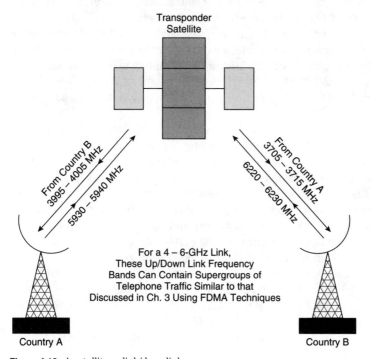

Figure 4.10 A satellite uplink/downlink.

they adapted well with satellite technology, and the INTELSAT system using time division multiple access (TDMA) was developed. Bit rates of 10 to 100 Mbps are used. The telephone industry uses a large number of satellite systems and has invested large amounts of money in their development.

To better understand satellite technology, a brief, simple discussion should be sufficient at present. Since the earth spins on an axis passing through the true north-south poles, a satellite placed in an equatorial orbit should spin at the same rate as the earth and hence will appear to stand still. There is no relative motion between the earth and the satellite. This, of course, was studied mathematically and proven by the space programs. Communication transponders powdered by rechargeable batteries charged by solar cells were placed on board small propulsion systems on the satellite. This was developed as a commercial use of the once military-controlled space program. The useful life of a satellite is limited by the number of charge/discharge cycles of the on-board battery system and the amount of on-board fuel. The propulsion system was used to make minor corrections in the satellites position and keep it on so-called station. Satellites are monitored by a telemetry system containing data on its operational status. When parameters change, that indicates an impending problem, and the uplink control can make a certain amount of corrective action.

Satellite useful life is about 10 years, although some satellite systems seem to be doing better. The need for satellite service has increased dramatically, and the orbit has become increasingly crowded. Satellite spacing is at present 3 degrees of longitude, and it could possibly be 2 degrees in the future. The close spacing of satellites makes it quite a problem for an earth-station antenna to distinguish which satellite is which. When the parabolic dish antennas and feed systems improved sufficiently, 3-degree spacing became possible.

When satellites are placed in orbit, relative signal strength contours are plotted on maps of the service area, one map for each satellite. This enables the signal to be estimated when an earth-based station is installed. When an earth station is to be installed, an accurate survey is necessary to determine the site latitude and longitude to the nearest minute. A topographical map can provide reasonably accurate results. However, the use of a theodolite (surveyor's transit) to site the North Star and an ephemeris (tables of celestial bodies in relation to points on the earth) will provide very accurate results. This procedure is best done by a registered surveyor.

Another method is to use a global-positioning satellite system, which gets data from a global-positioning satellite and computes the latitude and longitude at the receiver location. This method is used by maritime people and pleasure boaters. Once the site latitude and longitude coordinates are known, the look angles for the parabolic antenna can be calculated for the selected satellite position. An example of such a calculation is given in appendix K. Notice that for an equatorial orbit, the satellite only has a longitude value for its position over the earth's surface. Once the angles have been set using a surveyor's transit with compass for the azimuth angle, and an inclinometer for the elevation angle the satellite receivers should be activated. A channel that is active should be selected, and a power meter should be connected to the receiver's intermediate-frequency test port. Now the azimuth angle fine-adjustment is varied left to right while monitoring the instrument reading. This adjustment should be continued until the meter reading is maximized. This procedure is repeated for the elevation angle.

To measure the carrier-to-noise ratio in decibels, the maximum signal level for this channel should be recorded in dBm. Now the elevation angle is increased or decreased off the correct angle position to cause the signal to go into noise. Now the noise level is measured in dBm by the power meter. The difference in the signal level and noise level in dB is the carrier-to-noise level. This procedure can be repeated for each satellite channel, and the measurements can be recorded in a log book for comparison with later measurements. Adjacent satellite channels operate with different transmitting antenna polarization, which is either horizontal or vertical. Polarization adjustment should be made before making the carrier-to-noise ratio tests. The carrier-to-noise test procedure is shown in Figure 4.11.

There are many users of satellite technology in today's world. The telephone industry, the cable television industry, the federal government, and the television broadcasters are major users. Several satellite communications carriers have developed over the years and have financed and constructed satellite

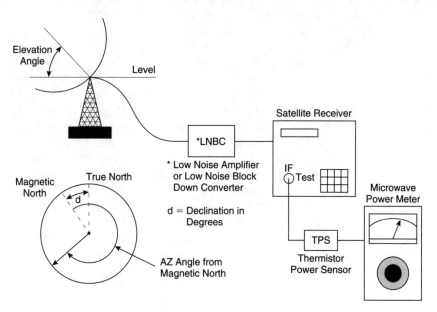

Figure 4.11 Satellite receiver C/N test.

systems that were launched by the National Aeronautic and Space Agency (NASA). Several corporations in the United States have taken part in the satellite program, either as a financial partner or a full working partner. This technology crosses over oceans, mountains, and countries where little or no communication facilities exist, and the technology will be around for a long time to come. As satellites in orbit fail due to power or fuel loss, new satellites are injected into available orbits. It was thought that the space shuttle could perform the refueling and battery replacement, thus recycling existing satellites by increasing their useful life. This is still a possibility.

4.1.3.6 Cellular and personal communications. Probably the most familiar and recent connection to the telephone network is through the cellular telephone and personal communication system (PCS). Most improvements in the telephone system are behind the scenes and out-of-sight. Therefore we often experience them but do not see them. Not so with the cellular telephone. Users of cellular telephones are often seen daily talking from their cars, offices, stores, schools, and even park benches. Some people wonder how they ever got along without a cellular phone. The benefits of such a system are easily recognized. People rescheduling appointments, emergency contact with medical, fire, and police, and location of material pickup and delivery are some of the obvious benefits. The pricing of such service evidently is low enough that it doesn't stop many people from using the service for everyday, mundane uses. At some business meetings, the participants are required to leave their cellular telephones in their offices so the meeting will not be interrupted by the ringing of cellular phones. So there are real advantages and real disadvantages.

The development of the cellular telephone was an outgrowth of the old frequency division multiplex (FDM) method of the mobile telephone system. When this type of radio telephone system went digital, it became known as frequency division multiple access (FDMA), where the speech was digitally encoded and modulated the RF carrier. Many developments have been made, and due to the chip manufacturers using very large scale integrated circuit (VLSI) technology, the portable telephones became more compact, with a longer battery life before recharging. Present-day fold-up cellular portable telephones are easily carried in a woman's purse, a man's inside coat pocket, or a person's briefcase.

The cellular concept means that a service area is divided into a geometric pattern of cells with a cell-central receiving-transmitting system placed in the cell's center. Omnidirectional or circular coverage takes place around each central cell station. Each cell is planned as an octagon shape, and the circumscribed circle provides some system cell overlap. This is shown in Figure 4.12. Even though the cells are planned as hexagonal, in practice, reflections from tall buildings in many city environments cause the actual cell coverage areas to change shape. Therefore cell overlap sizes are different for each cell.

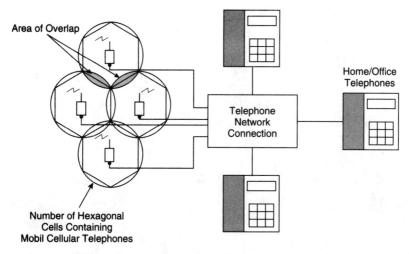

Figure 4.12 Cellular telephone system.

The cellular telephone is a low power transmit device usually in the 1- to 10-mW power range. When a cellular telephone moves through a cell and its receive/transmit signal gets weak near the cell outer edge, the condition of the receive signal by the base or central-cell station is sensed, and the base station of the entering cell takes over. This is known as a "hand-off" condition, where the leaving cell station transfers communication to the cell being entered. The hexagonal cell structure continues to many adjacent cells in a honeycomb pattern.

There are a limited number of frequencies allocated by the FCC to cellular telephone service, and as the size of the service area grows, some means has to be used to control the number of cells and frequencies. This means that frequencies can be reused if the cells are far enough apart. This frequency reuse method is very practical. Since the cellular telephones have basically a weak transmit power, their range is quite short and will not interfere with remote cells operating on the same frequency. This frequency scheme is shown in Figure 4.13. The number of cells and the number of frequencies used, determine the amount of cochannel (common channel) interference that can accumulate and impair communications.

There are tradeoffs in cell pattern and the number of available frequencies. Cell sizes often have a radius of several kilometers, and a cell with a radius of less than 1 kilometer is referred to as a microcell. Cellular telephones working in larger cells require the higher output of about 10 mW to work properly. The microcell concept is often associated with the Personal Communication Network (PCN), which uses the smallest portable cellular telephones. The high operating frequency requires a very short antenna

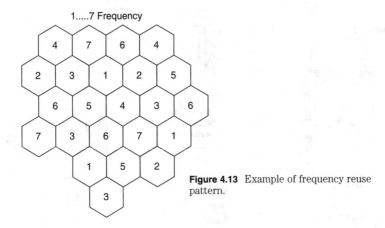

Figure 4.13 Example of frequency reuse pattern.

and so allows the pocket-type cellular telephones to be small and compact.

Presently in the United States, Europe, and Japan, cellular telephone service is operated in the 825-to 950-MHz frequency band. Even though the frequency bands are close, there are major differences in the upstream (mobile) and the downstream (base) channel plans between the United States, European, and Japanese cellular systems. Only in Europe does a problem occur due to differences in systems in adjacent countries. A cellular telephone subscriber will be confined to using the cellular phone in the country of system origin and will not be able to go from country to country. The only problem with incompatibility we, in the United States, might have is that we will not be able to market our equipment in other countries unless the products are specifically manufactured for export.

The allocated cellular channel plan is shown in Figure 4.14. Portable cellular telephone systems operate over the air between the telephone and the base station, and some of the radio frequency scanners sold in many of the electronic outlet stores are able to monitor some of the cellular telephone systems. Therefore some cellular telephone conversations have been reportedly overheard by some of the scanners. This can

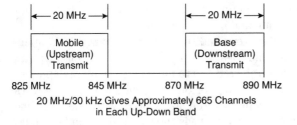

Figure 4.14 Cellular telephone channel plan.

become a problem for certain users of the cellular telephone systems. Many developments in the cellular systems are forthcoming, and an all-digital system with digitized speech and digital message scrambling will solve most of the problems. This is an important and popular feed into the wire telephone domain. Thus computer modems operating in conjunction with the cellular telephone system make transfer of digital data and fax data from remote sites possible. Theoretically one could fax documents from a park bench if desired.

Recently the FCC has conducted a so-called auction of frequency bands for cellular service. Advancements in microchip sets and digital technology will no doubt take place to further this section of the telecommunications industry Cell size will no doubt become smaller, and some cable television operators have proposed using their return signal path as an upstream connecting path into the cell central base station. Rest assured. We can expect a lot more development to take place with cellular telephone and personal communication systems.

4.2 Cable Television Signals/Interconnects

Cable television systems essentially distribute television RF signals containing television programming to subscribers. This is actually a TV delivery service where a large number of television programs are offered in packages to the paying subscribers. The cable system started as a simple extension for the television broadcast stations to outlying remote areas with poor reception, but it developed into a highly sophisticated cable system carrying a huge amount of television services.

In the United States, the television signal was determined by the National Television Systems Committee (NTSC) and was for the early monochrome-only television available at that time. When color television was being introduced, the signal had to be compatible with the existing monochrome receivers. This was accomplished by some minor changes in the monochrome signal needed to accommodate the added color information. The scanning standards for the NTSC color television broadcast signal is 59.94 fields per second and 29.97 frames per second. This is one-half of the field rate, since there are two fields per frame.

The television signal is made from horizontal (across-screen) lines beginning from left to right and progressing down the screen top to bottom, drawing what is known as a raster. Originally the field rate for monochrome was 60 per second, with a frame rate of 30 per second, and the field rate was at the same frequency as commercial electric power. Each field interlaced with each other which reduced the line effect on the picture.

The color television baseband video signal is very complicated and carries an enormous amount of information. Picture synchronizing, color, brightness, compatible stereo/mono sound, digital closed-captioned signals, and test signals make up the information carried in the television broadcast signal. This baseband signal modulates an RF carrier using what is known as vestigial sideband amplitude modulation (VSBAM) for the video signal and frequency modulation (FM) for the audio signal.

The bandwidth for one television channel is 6 MHz wide and is shown in Figure 4.15 for television channel 3. At first, television broadcasting only had 12 available channels [channel (2–13)], with the first 5 channels at what is known as low VHF., 54 to 88 MHz for channels 2–6. The other 7 channels occupied the high VHF band at 174 to 216 MHz for channels 7–13. Since more always seems to be better, the public demand for more television broadcasting stations caused opening up the ultra high frequency (UHF) band. At 470–890 MHz, this band gave channels 14–83 for an additional 70 channels. Now television sets required a VHF dial and a UHF dial to enable the set to tune to all these stations. Broadcast stations operating in the UHF band required much more transmitting power in order for the signal to arrive at the receiving antenna with sufficient strength to compete with the VHF stations. This was an added expense for UHF broadcasters. Hence not many UHF stations came to exist. In most major cities, more UHF stations exist.

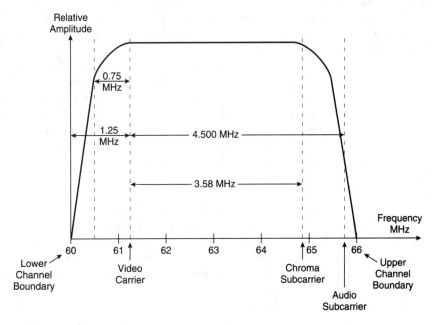

Figure 4.15 Television channel 3 band allocation.

4.2.1 Basic television services

As stated before, the meat and potatoes of the cable television industry is the delivery of television programming to paying subscribers. The sole source, at one time, was the carriage of off-air television broadcast stations of the VHF type. The cable system was essentially connected from the utility-pole-mounted plant to the subscriber's television set antenna terminals via a drop wire. The coaxial cable carrying the television broadcast channels has a loss characteristic that increases with frequency. Thus the transmission of UHF stations was impossible for on-channel operation. Since the cable system is essentially sealed, the FCC allowed cable operators to use the frequency bands occupied by other off-air radio services. These services were located between channel 6 and 7 in the VHF band, which consisted of the FM band (88–108 MHz) and the aircraft safety voice and navigational aid frequencies 108–174 MHz. Many cable operators elected to carry the FM stereo radio band, offering it to their subscribers as a service at usually a small additional cost. The remainder of the frequency band allowed 9 additional television channels. Therefore UHF stations were converted to these channels, transmitted to subscribers through the cable system, and converted by a separate set-top converter back to actual UHF channels. These were then selected by the subscriber's UHF channel selector. Now the cable system could offer more stations, thus making it more desirable for subscribers to hook up.

4.2.1.1 Off-air television reception. Cable television signals originate at some central point called the head-end, where they are processed and adjusted before entering the cable distribution system. As previously discussed, the cable system consists of two classes of cable, the trunk cable and the distribution cable. The trunk cable transports the signal to the extremities of the service area through a cascade of repeater amplifiers properly spaced along the system branches. The distribution cable is fed from the trunk cable system from bridging or isolation amplifiers and contains the subscriber service tap devices. In heavily populated areas, the distribution cable runs alongside the trunk cable, both lashed to the same aerial steel supporting strand.

The main source of the off-air television broadcast stations are from tower-mounted antenna arrays. In many instances, one antenna element receives one or more stations for a given direction. The antennas are connected to electronic processing equipment in a building or enclosure located near the base of the tower. Each channel is individually processed, and the video and audio carrier levels are adjusted according to cable transmission standards.

Tower height is determined by the number of receiving antennas, the relative signal strength of the received stations, and local-area height restrictions. If the tower is higher than 200 feet or an airport runway approach is nearby, Federal Aviation Administration rules might require the tower to be appropriately lighted with red warning lights. The top light is a flashing light. A sketch of a tower and equipment diagram for off-air reception is shown in Figure 4.16.

Off-air reception for a television broadcast station uses an array of antennas that will provide sufficient signal strength to produce the required carrier-to-noise ratio (CNR). Signals with appropriate CNR produce high quality video pictures. The picture will never be better than at the head-end. Each channel is individually processed by either a strip amplifier or a heterodyne processor. Early systems used strip amplifiers that just passed through the selected channel. The level of the video and audio carrier was adjustable, and the audio carrier level could be separately set relative to the video carrier. The heterodyne signal processor tunes or selects the desired channel and separates the intermediate frequencies of the video and audio carriers where they are adjusted in level. Then these intermediate frequency (IF) audio and video carriers are converted to the desired cable channel.

Heterodyne signal processors have the capability of automatic gain control (AGC), and the output channel can be any desired cable channel, not just the same as the input channel. Strip amplifiers often had AGC capability, but operation required the output channel to be the same as the input channel since they were essentially amplifiers. The block diagram for these two pieces of equipment is shown in Figure 4.17. Present-day heterodyne processors have a great many desirable features. Alternate sources of intermediate frequency (IF) signal can be automatically injected. Such signals contain alternate programming sources. An alternate source of audio

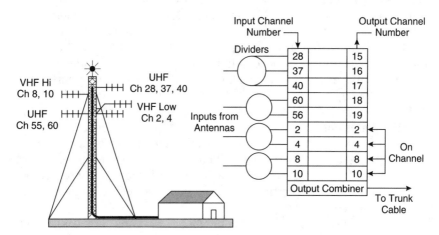

Figure 4.16 Off-air television receiving section of a cable television head-end.

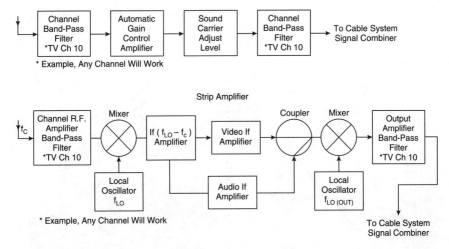

Figure 4.17 Heterodyne signal processor.

can act as an override alert for a disaster warning system, often required by cable licensing authorities. The heterodyne signal processor is used by many cable operators mainly because it is extremely reliable and very versatile. It enables cable operators to receive a channel on the broadcast frequency and place it on the cable on another frequency. This is very necessary for UHF stations where they are received at the high UHF band and placed on a lower-frequency cable channel. There has to be one signal processor for each off-air channel.

4.2.1.2 CARS band microwave links.

Microwave radio techniques have been used by many cable operators to distribute cable programming to many HUB sites. This method is extremely useful when mountainous terrain makes off-air reception difficult. Signals from off-air are received in an area of good reception, often a mountaintop, and distributed to other sites via line-of-sight (LOS) microwave radio. Often these hub sites served small communities located in valleys. This concept is shown in Figure 4.18.

In practice, microwave links transmit to many remote sites, making this a cost-saving and effective method. In many cases this might be the only viable method. This method does not require any connecting roads or poles in between the transmit/receive path to bring service to outlying areas. This relay service used by cable operators is known as cable antenna relay service (CARS) and operates in the 12–13-GHz band. Operation in this band is often referred to as AML, which stands for amplitude modulated link. This service is a licensed service, and either the cable operator has to secure a license to transmit, or, where available, the service can be provided by a microwave AML common carrier who carries the license and distributes

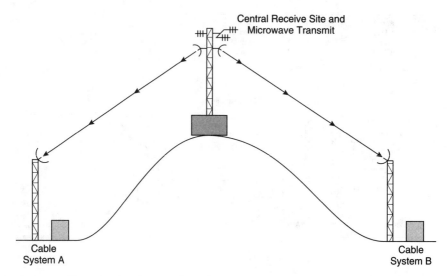

Figure 4.18 Example of a microwave line-of-sight relay system.

service to several cable operators along the system path. Each AML channel is 6 MHz wide to accommodate the standard television channel.

Television programming appearing on standard television broadcasting channels are frequency multiplied by heterodyning to microwave frequencies in the 12–136 GHz band. For example, take television channel 10 with a frequency band of 192 to 198 MHz and a microwave local oscillator operating at 12646.5 MHz. The sum frequency from the heterodyne process produces a frequency band of 12838.5 to 12844.5 MHz in the microwave region. Several consecutive, standard VHF channels can be transmitted by one microwave transmitter. If more channels are required, several more can be grouped and transmitted by another microwave transmitter. Microwave transmitters are often joined in a combining network and share a common parabolic transmitting antenna.

The reverse process takes place at the receiving sites, where the microwave receivers convert the 12–13-GHz band to the groups of television broadcast channel frequencies. These groups of received television channels are then combined and fed to the connecting cable television plant. Often, CARS-band channels are used to provide studio-to-transmitter links (STL) for the television broadcast stations. Most studios are located in a city environment, and the transmitters are located in a surrounding remote area. The studio-originated video and audio signal is microwaved out to the transmitter site. Monitoring and test signals are usually sent back to the studio via a reverse-direction microwave link, which is used to provide data for the transmitter operation.

Some CARS-band microwave links use frequency modulation (FM), which operates with two bandwidth options. One option requires 25 MHz, the other 12.5 MHz. This depends on the application. Baseband signals can be either baseband video (NTSC) or digital bit streams carrying computer data and/or digital telephone traffic. As far as cable operators are concerned, use of baseband signaling is confined to distributing television programming to remote hub sites.

4.2.1.3 Satellite receiving stations. Cable television operators were quick to use the satellite system spinoff from space exploration. Home Box Office (HBO) was one of the first of the premium programming suppliers to use satellite service. WTBS, the forerunner to the vast Turner corporation television group, was one of the first television broadcast stations to go on the satellite system. Such stations became known as superstations.

Early cable systems used very expensive, large receiving antennas some as large as 10 m (33 ft), accompanied with low-noise amplifiers at the antenna. Connecting cable was large in diameter (⅝ to 1 inch) for low loss and was pressurized with dry air to keep out moisture. Each receiver operated on a specified channel, where FM modulation was used. The receiver output was baseband (NTSC) video and mono audio, which then had to be modulated in standard VSBAM television manner to feed into the cable system.

Television channels received and modulated in this manner were combined with the off-air processed signals to make up the cable television service. Signal security of the premium movie service (HBO) was done by *signal traps*. A signal trap is a deep notch filter that deletes the premium signal from a subscriber tap port. Nonsubscribers to the premium service had to have a trap to stop the signal from entering the premises. An expensive proposition it was, employing many traps and personnel time to install them. If a subscriber decided to take the service, a technician and a truck-roll was required to remove the trap from the line. Many techniques have been used by cable operators to control the signal security, allowing it for paid subscribers and eliminating it from just basic subscribers.

Once the satellite system matured, many satellites supplied a large amount of available programming to cable operators. As antennas became smaller, block down converters with built-in low-noise amplifiers became available, and improved receivers were developed, system costs decreased.

Some individuals with money bought their own receiving systems and were obtaining the service for free. Program suppliers, mostly the sports and movie suppliers, started to scramble or encode their signals. Now the cable operators were required to put in sophisticated decoding or unscrambling equipment. Since this seemed like a good method of providing signal security, cable operators rescrambled the signal to subscribers. The decoding or unscrambling method took place in the set-top cable converter.

A diagram of a present-day satellite receiving system is shown in Figure 4.19. The 3.7- to 4.2-GHz band is split into 24 channels 20 MHz apart. Alternately numbered channels are horizontally polarized. For example, even-numbered channels are vertically polarized, and odd-numbered channels are horizontally polarized. The horizontal and vertically polarized channels are 90 degrees shifted in phase. There is now 40 MHz between each adjacent, horizontally polarized channel and 40 MHz between adjacent, vertically polarized channels. This separation in frequency and in polarization provides proper signal isolation from adjacent channels. Satellite service is extremely important to cable operators and generally provides the bulk of the service other than the off-air television stations.

4.2.2 Reverse/upstream systems

Many cable systems were required by the licensing or franchising authorities to install reverse or upstream transmission capability. Since the licensing or franchising authorities were appointed by the local municipal governments, local interests were at the heart of the matter. Cable operators were well aware of the unused frequency band below channel 2, and equipment manufacturers developed a method of upstream (reverse) transmission in the 5–30-MHz band. Four-and-a-half 6-MHz video channels could be used in this band.

The frequency band between 30 MHz and the lower boundary of the first downstream channel (TV channel 2) was needed to separate the upstream band from the downstream band. This was called the crossover band and was separated by a filter called a diplex filter.

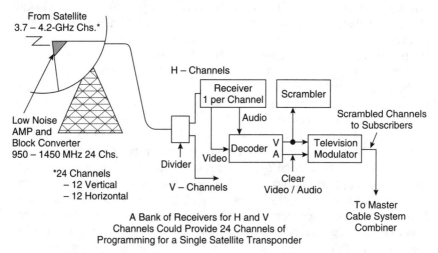

Figure 4.19 Cable television satellite receiving system.

This frequency plan was termed the *subsplit reverse specification.* Reverse amplifier modules were made to fit into available downstream amplifier housings. The diplex filters appeared in the upstream amplifier module input port, since the cable loss was less in the reverse band and the amplifier was not needed at each downstream amplifier location.

4.2.2.1 Single cable systems. This method worked quite well, and many systems only actually used a couple of upstream channels, usually from a high school and/or town hall. However, some municipalities had grandiose ideas and required the cable operator to provide more upstream channels. The licensing or franchising process usually required hearings where cable operators could explain the options and costs. As cable system equipment improved, the downstream channel capacity was increased greatly. Present-day systems often have an upper frequency limit of 600 MHz, and 750 MHz to 1 GHz is being proposed. A 600-MHz system has a downstream channel capacity of about 86 channels.

A system called the midsplit system has an upstream band of 5–112 MHz and a downstream beginning at 150 MHz, on up to the upper frequency bound. This 5–112-MHz band typically yielded 18 upstream television channels that were 6 MHz wide. When the upper limit was 300 MHz, the 150 MHz of the downstream (300–150=150) gave 25 channels and hence was close enough to be called midsplit. The 18 upstream channel capacity was found to be more than adequate for many cases. However, the cable manufacturers prepared for the worst and made equipment for the system called the high-split. This plan had split band that was increased upward in frequency. The upstream band became 5–174 MHz, and the downstream began at 234 MHz on upward. Since the upper frequency limit was pushed to 550 MHz, the downstream channel capacity was 52 channels and the upstream was 28 television channels. This was accomplished by a single cable system with the upstream/downstream channels separated by diplex filters.

The subsplit, midsplit, and high-split plans are shown in Figure 4.20. In practice the subsplit method was the most common, and there are only relatively few midsplit and high-split systems. A typical amplifier configuration for the subsplit reverse case is shown in Figure 4.21.

4.2.2.2 Dual cable systems. A separate cable system that was used for feeding television signals from remote locations back to the head-end site was usually constructed in heavily populated urban areas. This type of system was constructed where there was a requirement for many reverse channels. This type of construction was referred to as a *dual cable system,* where there was the downstream subscriber cable system and the upstream separate reverse cable system.

Cable operators who either proposed such plans or were required to by the licensing authorities found that there were both benefits and problems.

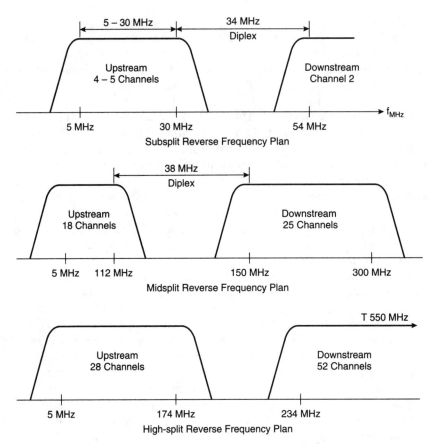

Figure 4.20 Single cable upstream/downstream transmission methods.

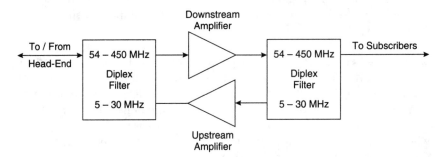

Figure 4.21 Example of a subsplit reverse amplifier configuration.

The benefits were that television services on the upstream or in-bound cable could be dropped along the way to other locations. For example, one school could originate a program and drop it off to a another school along the way. Some elaborate dual-cable systems had an inbound and outbound system, often called an *institutional network* (INET), which provided video, audio, and digital data services to municipal buildings. Often this type of system was entirely separate and in some cases did not pass through the head-end. However, most of them were connected to the head-end.

This type of system is illustrated in Figure 4.22. The separate cable system can carry standard television service on ordinary television channels. Therefore the schools can use regular television sets, cameras, and equipment. School A could distribute programming on television channel 4 to school B, the town hall, and the cable head-end. If the program was to be sent to the cable subscribers, channel 4 could be converted to any available

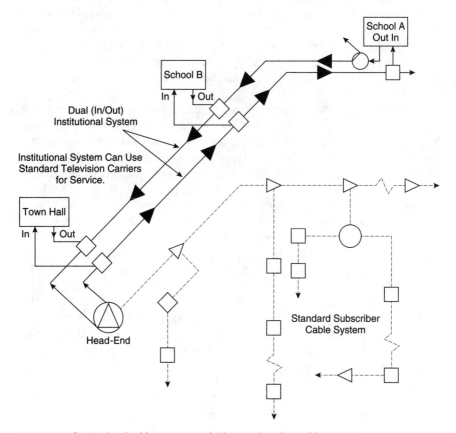

Figure 4.22 Institutional cable system overlaid on a subscriber cable system.

cable channel that the cable operator had available. This type of system could carry a full complement of 60 channels or more in both directions. For most municipalities, this was considered overkill, and in most cases only a few channels in each direction were used. This type of system was also expensive to power and maintain, the cost of which was usually borne by the cable operation.

4.2.2.3 Reverse network topology. The network topology developed for the cable television industry was of the tree-and-branch variety. This worked well because the purpose was to download television programming in sufficient variety and quantity to paying subscribers. From what started as a simple extension of a superantenna known as community antenna television (CATV), a highly developed cable television industry was formed. Television programming in the form of the NTSC video signal was assembled at the central head-end and placed on television carriers. Since some of these carriers were not selectable by some of the subscribers' television sets (noncable ready) the cable operator provided some form of set-top converter that would place cable channels on a selectable channel. For the premium services, signal scrambling methods were developed that made the premium (movie) services unviewable on a nonpremium channel subscriber's set.

The unscrambling task for premium channel subscribers took place in the set-top converter. Basically, cable television systems are essentially a one-way delivery system to cable subscribers. However, extra revenue in the form of alarm monitoring services, smart set-top converter boxes, pay-per-view services, and utility meter-reading services caused the cable industry to study the reverse or upstream capabilities previously discussed. These reverse or upstream frequency bands for the single cable system were structured to provide upstream service. The reverse network for tree-branch topology suggests that the outer, small branches feeding back to the head-end will cause a large buildup of noise. When some systems installed reverse amplifiers, the noise buildup was indeed confirmed.

Therefore, to make the reverse system operational, only the path necessary for transmission was activated back to the head-end. To activate the switch, a scheme for a smart switch was developed. A computer system would download digital data usually in the upper FM band, which would address the switches required to activate the return path. Smart converter boxes could be sent instructions conforming to subscribers' orders via this computer on the downloaded data channel. The addressed converters could answer with subscriber request or orders for service by placing digital data on the reverse path. When this reverse path was active, the digital data could enter the service computer. Pay-per-view service is done this way on some cable systems. It must be remembered that cable systems today carry NTSC-type VSB modulated carriers. However, cable amplifier cascades can also carry large amounts of digital data as performed by the telephone industry.

In recent years, cable television technology has found that fiber-optic methods are extremely useful and practical. Simply stated, in the normal cable television networks, the trunk cable acts as the main transport and contains no subscriber taps. As previously discussed, the distribution cable system is bridged from the trunk through an isolation amplifier (bridging amplifier) to the distribution cable containing subscriber taps and short (2–3) amplifier cascades of distribution amplifiers. Cable operators were quick to realize that fiber-optic cable could be used to replace the trunk cable. Since the loss of fiber-optic cable is so much less than coaxial cable, no repeaters are needed. Therefore the fiber-optic cable would be entirely passive from the head-end to the receiver at the distribution point.

Many cable operators are presently installing fiber-optic cable for various upgrade and rebuild applications. One such trunk scheme is illustrated in Figure 4.23. Fiber-optic cable use in cable television systems serve a variety of purposes. One fiber can essentially completely carry most cable television systems services, and in commercially available fiber-optic cables, several fibers are standard. Fiber-optic cables are available in two to several hundred per cable. Eight, 16, and 24 fibers per cable are most common for cable television applications. For example, when cable operators install 16-fiber cable and use 6 fibers, they leave 10 for upstream or other uses. These other uses could include meter reading, interactive television service, digi-

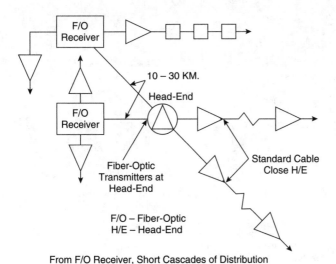

From F/O Receiver, Short Cascades of Distribution
Amplifiers and Taps Serve Subscribers. At or Near
the H/E, Short Cascades Serve Subscribers.

Figure 4.23 Fiber-optic trunk (backbone) method.

tal services such as information and/or games, music, or leased lines to other carriers.

The fiber-optic transmitters are usually lasers whose light output power is modulated at the signaling rate. The receiver usually consists of optically sensitive PIN diodes, which convert the varying optical energy transmitted to electrical energy output. For multichannel cable television service, the frequency spectrum of the RF carrier is usually grouped into, for example, 12 carriers each to drive a fiber-optic transmitter. At the receiver, the light energy is converted back to the 12 carriers, which is then combined with the next group of 12, etc., until the cable service is assembled in a combining network connected to the distribution amplifier cascade. With the development of high-output optical transmitters and sensitive receivers, in many instances one fiber can transmit the entire cable system service from 15 to 30 km. An example of a case where standard AM VSB signals are transmitted optically is shown in Figure 4.24.

4.2.3 Radio and miscellaneous services

Cable television systems carry several types of signals besides standard television programming. Some of these signals, such as FM stereo, background music service, or premium-music radio, are offered to subscribers either as an extra-cost service or are included with basic service. Other signals might be control signals for the aforementioned bridger-leg reverse switching or to control premium-channel programming. Each of these types of signals is markedly different from the television signals.

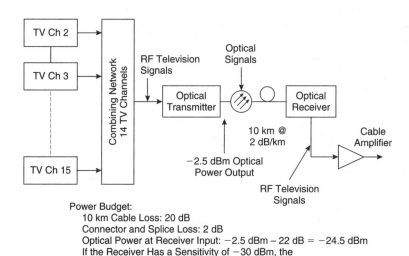

Figure 4.24 AM fiber-optic transmission of cable television carriers.

As previously discussed, premium-programming security and subscriber authorization control was accomplished in the set-top converter. The television standard NTSC video signal was relieved of its horizontal synchronizing pulses, thus making it impossible for the subscriber's television set to hold the picture still. A subscriber who desired to purchase a premium channel had the cable company install a programmable read only memory (PROM) installed in the set-top converter. This chip authorized the circuits in the set-top converter to restore the synchronizing pulses, thus allowing the television set to produce a picture. Later the PROM was replaced by addressable chip sets, where each converter had a digital address and authorization memory. Then the subscriber desiring service would contact the cable company, who through a computer terminal would address the converter and then send the authorization code. This downstream converter control chemical FSK-modulated an RF carrier, usually located at the top of the FM stereo radio band. This carrier with addressable data gave the cable operator some experience transmitting digital data on the cable system.

4.2.3.1 FM broadcast service. FM radio services have been carried on many cable services for many years. Since the FM radio band happened to be located between television channels 6 and 7, this made carriage of FM radio service a natural. Early cable systems installed an FM antenna on their receiving tower, and using an FM band amplifier applied the output to the cable system. Strong FM signals were attenuated by series-tuned inductive-capacitive (LC) circuits, and the amplifier gain was adjusted for proper signal levels. In many systems this method provided satisfactory results. For many urban and suburban areas with FM signals of varying strength, a more elegant treatment was required. Single-channel processors were used, which received the FM signal, converted it to the standard 10.7 MHz intermediate frequency (IF) with automatic gain control AGC circuits, and then converted it back to a standard FM channel frequency. The output channel could be on the same frequency as the input channel (on frequency) or converted to another standard FM frequency (off frequency). Now the FM radio service appearing on such a cable system had the FM carriers neatly spaced and all at the same level. Subscribers could easily tune through the FM band, finding stations with equal signal levels. Maintaining the proper FM channel bandwidth also allowed these FM stations carrying background music services to pass through the FM cable service.

4.2.3.2 Digital radio service. High-quality audio services available through some of the satellite is one of the latest offerings by the cable television industry. The quality is similar to that obtained on compact disc. Subscribers to such services play this through their stereo systems for either personal entertainment or background music service. This type of service is essentially digital audio, and either subscriber or head-end equipment is necessary to accept the cable signal and convert it to normal baseband left-and-right-channel audio signals. Thus another piece of add-on electronic equipment is needed by the subscriber to process this service if the cable operator elects to transmit this signal as a modulated RF carrier. Some cable operators might decode this service at the head-end and transmit it on the cable as an optional premium audio service on an FM carrier. Such service already is or is soon to be offered to direct broadcast television (DBS) subscribers.

4.2.3.3 Digital television services. In the not-too-distant future, digital video signals will become available to the cable operators. Presently, digital transmission of video programming is available directly to paying subscribers through the Direct Broadcast Satellite (DBS) system. Microwave carriers in the Ku band (12 GHz) transmit digital-coded program packages to subscribers. Small (18-inch) offset-feed receiving antennas and selectable receivers make up the subscriber equipment package, which costs typically under 1000 dollars. Installation is simple, and antenna pointing is aided by entering the subscriber's zip code into the receiver, which then displays printing information. System picture quality is similar to that obtained in video CDs, which also use digital video signals. Programming-package costs are considered competitive with cable television prices. However, in many cases users can get more service from cable operators and have no electronic equipment to maintain.

Cable operators might find some problems coping with digital video, mainly because not much is known about how existing satellite equipment will react to digital signals. Possibly some pretesting could be done to see how equipment such as the low noise amplifier (LNA) or low noise block converter (LNBC) and/or satellite channel receivers operate with digital signals.

Since FM-type modulation is often used, the receiver de-emphasis network might need to be replaced for digital signals.

Digitally coded, full-motion video takes about 45 Mbps in its normal form. With data rates this high for one video channel, the cable television bandwidth would be quickly eaten up. Much information in a typical video scene does not change much from frame to frame. If such redundant information is transmitted once and only the portion of the scene that changes is transmitted, much information could be deleted from transmission. This process, known as *video compression*, has been studied extensively by the Motion Picture Experts Group (MPEG) and it proposes two standards, known as MPEG-1 and MPEG-2. MPEG-1 is full-motion, compressed video resulting in a bit rate of 1.5 Mbps. Although this is workable for some applications, there is enough picture degradation to warrant a higher standard, known as MPEG-2. This standard typically uses about 4 Mbps for full-motion NTSC video and 10 Mbps for the proposed high-definition television (HDTV) system. The solid-state chip manufacturers have produced the necessary chip sets to perform the digital compression/decompression tasks. How cable operators handle such digital techniques remains to be resolved. However, digital video will certainly be traveling on the digitally oriented information highway. Compressing the digital video decreases the bit rate from the uncompressed state about 10 to 1.

4.3 Private LAN/WAN Service

The need to connect personal computer workstations to mass storage equipment and shared printers was one of the first of the requirements for early local area networks (LANs). As previously discussed, available wiring used by other communications facilities was used first. Since most of the requirements were in the same office area or building, the connecting wires were relatively short. The connections to mass storage equipment had to be managed by microprocessor equipment called a *file server*. Equipment, the accompanying software embedded in the file server itself, and an installed internal accessory circuit board or external modem allowed for orderly, filed tasks to be carried out by the workstations. Similar data communications between shared printing equipment allowed for printing to take place in turn in an orderly fashion between workstations. As systems became larger, the need for workstations to communicate with one another developed.

The communication system distances increased, often beyond buildings, and thus the LAN in a campus environment developed. Presently, LANs belonging to the same company are interconnected across the country by several means. A satellite system referred to as Very Small Aperture Satellite (VSAT) is quite commonly used by the banking industry, the automotive manufacturers, and the credit and financial community. Also, the telephone industry and other common message carriers provide either hard-wire leased lines, coaxial cable channels, microwave radio links, or fiber-optic cable. The method of choice is formed by the user's requirements, availability of the communication facilities, and, of course, the cost.

4.3.1 Baseband digital signals

The data generated by computer equipment appears as a train of pulses that describes the information bits of the two digits 0 and 1. If the pulse level that corresponds to a binary 1 returns to 0 volts before going to a 0 or 1 again, this is referred to as a return to zero (RZ) type. If the level stays for another clock-timing period for another 1 in the sequence, this is known as nonreturn to zero (NRZ). If equipment corresponding to one type is going to be used with equipment that uses the other type, clearly some interfacing is needed. Also, the pulses, if placed on a common bus, are to be connected to many pieces of equipment through long cable runs; circuit loading and reflections due to impedance mismatching can distort the pulses to a point where data is not transmitted error free.

Thus early on, the modem was developed to overcome this problem.

4.3.1.1 Types of system network topologies. Early computer modems operated in a local area type of environment, and since the modulation and signaling schemes were compatible with the telephone service, the interconnection between LANs was possible. These early modems were quite slow by today's standards, but they fulfilled a need. Modem standards were developed by the telephone industry according to Bell System standards, the Electronic Industry Association (EIA), and the international CCITT. Essentially, early modems selected one of two tones (frequencies) to describe the binary 1s and 0s, and was referred to as *frequency shift keying* (FSK). Data transmitted in this manner appeared on the cables as a couple of frequencies being selected. These were sinusoidal tones, and the signaling rate of 300 bits per second was common. Naturally high data rates required modems to use different keying methods.

Over the years the types of modulation included frequency-division modulation and various phase-shift types of modulation. One method uses a four-phase system, which is used to describe all possible values of two binary bits 99, 01, 10, 11 accounting for four binary numbers of two bits. When using eight phases, eight binary numbers can be realized. These types of modems had bit transfer rates of 1200 to 4800 bps. Higher-speed modems employed a combination of phase and amplitude modulation called *quadrature amplitude modulation* (QAM). Modems operating in this manner obtain bit transfer rates of 9600–28800 bps. Signal quality had to be preserved on the connecting cable systems so errors were minimized. The chip-set manufacturers have developed the necessary integrated circuits to the point that modem cards can be installed directly into the workstations.

4.3.1.2 Network control methods. The network control of data communication systems requires both software and hardware. The software consists of the programming protocols used by the communication interface circuits in each workstation and peripheral equipment such as mass storage and shared equipment. Much depends on the network topology and the network access control methods. Recall that the Ethernet system uses the Carrier Sense Multiple Access Collision Detection (CSMA/CD) protocol. Also, other networks using the token passing, token ring, or token bus conform to the Carrier Sense Multiple Access Collision Avoidance (CSMA/CA) protocol.

Present LAN implementations are of the upgraded Ethernet and/or Fast Ethernet, which support a maximum bit rate of 100 Mbps. An Ethernet data transceiver printed circuit board is required to be installed in workstations connected to an Ethernet LAN. A coaxial cable system is connected to the data transceiver. Manchester digital encoding is used where bit transitions also act as clocking pulses. This is shown in Figure 4.25. The Ethernet LAN implementation conforms to IEEE 802.3 standard. Data packetizing/depacketizing, data link management, and data encoding/decoding are performed by the data transceiver circuits. Ethernet LAN implementation transferring data at 10 Mbps can be transmitted over unshielded twisted dual-pair cable or coaxial cable. When standard unshielded twisted-pair telephone wire is used as the transmission medium at 10 Mbps, it is referred to as a 10-base T system. Transmission distances for a 10-base T system is 100 meters using number 24 AWG gauge wire.

Fast Ethernet systems operating at 100 Mbps use either coaxial cables for shorter transmission distances or fiber-optic methods. Standard cabling and connectors for Ethernet type LAN implementations, along with a variety of control circuitry and software products, support the use of Ethernet-based LAN technology.

Many other hard-wired or cable-type LAN implementations were developed along the way, and some are still in use today. International Business Machine (IBM) Corporation promoted several LAN implementations. AT&T, 3 Com, and Novell also produced various LAN architectures with available software and hardware. Various network topologies such as token passing, ring-and-bus methods, and star networks were used. The IBM-compatible personal computers (PCs) were used as workstations due to their modular construction and because they were easily adaptable to almost any type of LAN system.

4.3.2 Broadband computer data systems

Many people involved with data and computer system communications first looked at other available communication systems to see what techniques would prove useful. Use of the public telephone system (PTS) and telephone-type technology and equipment found its way into the LAN industry. As higher data rates were required, other technologies were explored. The cable

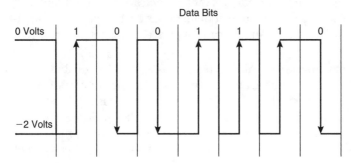

Figure 4.25 Manchester-type coding.

television industry, with highly developed coaxial cable and repeater amplifiers, looked to be an attractive option. This approach was used to develop
the larger-scale LAN systems and the wide area network (WAN) systems. The
cable television industry was already experiencing the requirements for single-cable, two-way service, and dual- or multiple-cable institutional networks
(INET). So cable television technology was used to develop what is known as
broadband LAN-type implementations.

4.3.2.1 System types and network topologies. From previous work, a single-cable coaxial-cable system used by cable television operators provided
service in both the forward and reverse directions. For the subsplit reverse
type, the forward (downstream) frequency band was from 55 MHz to the
upper frequency limit for the amplifier cascade. For example, 55 to 300
MHz systems gave 35 forward channels, and for 450 MHz, the upper limit
was 52 channels. The reverse or upstream band was from 5–30 MHz, allowing 4½ television channels. Thus there is more downstream channel capacity than upstream channel capacity. For cable television systems where the
most need is for downstream distribution of multiple television channels,
this is fine, but for an equal upstream/downstream data path requirements,
this is poor.

Also available is a midsplit system where the split isn't really in the middle of the frequency band. A high-split system reverse band of 5–174 MHz
gives 28 channels up and has a forward band of 234.400 MHz, which gives
about 28 television channels downstream and would work as a single cable
LAN/WAN.

This case gives an upstream bandwidth of 174–5=169 MHz and a downstream band of 40–234=166 MHz, all using standard available cable television components. Since the system power supplies installed along the cable
route power both the forward and reverse systems, this limits the number
of power supplies.

This type of LAN architecture by its nature is a common bus or ring-type
network that could use the available token-passing methods. The data bit
rates can be used to modulate standard television carriers with bit rates of
1–2 Mbps each. Various other modulation methods such as amplitude modulation phase shift keying (AMPSK) can support 10 Mbps rates per carrier.

4.3.2.2 Dual cable networks. Although far more expensive, dual or multiple cable systems can be used to support dual ring topology or dual bus
topology at high data transfer rates in both directions. This cable architecture is similar to that discussed in section 4.2.2 and shown as an institution
section in Figure 4.22. This technique operated with RF carriers going in
each direction carrying digital data. Since the shielded coaxial cable kept
electrical noise to a minimum, this type of LAN found use on factory floors.
The General Motors Corporation adopted this technique over Ethernet and

used it effectively on the noisy automobile manufacturing environment. This type of system is referred to as Manufacturing Automation Protocol (MAP), as specified by IEEE 802.7 standard.

This type of LAN used cable television-type technology in which the manufacturing processes used computer-controlled machinery and robotics technology. A protocol known as Technical Office Protocol (TOP) was used to extend this type of system to the office environment. The combination is known as MAP/TOP. Typically, a high-split single cable system with equal inbound and outbound frequency bands were used. This type of LAN easily supported voice, video, and data bidirectional communications. Data channels did not require any common synchronizing schemes, and so they could operate either synchronously or asynchronously. System network topology could be either tree, ring, or star type. For longer cable runs to outer buildings, system trunk cables connecting to the distribution cable system similar to the cable television trunk feeder design are often used.

Cable television equipment manufacturers were quick to pick up on a new business. Different value taps and cable equalizers were manufactured for the new LAN systems. Automatic gain controlled (AGC) amplifiers and uninterruptible power supplies provided the required supersystem reliability. Data transceivers were developed to act as the interface between the computer workstations and the data communication medium. Standard television carriers could also be carried on this system, providing video plant process monitoring, security surveillance, or video training. A design example is discussed in appendix M.

4.3.3 Fiber-optic computer data systems

The development of fiber-optic cable and the accompanying fiber-optic transmitting and receiving devices is revolutionizing the communications industry. The bandwidth of the fiber-optic cable is extraordinary, and the limiting factor is essentially the transmitting (laser) electronics and the receiving (photo diode) electronics. It was once thought that metallic wires were all that was needed to transport electrical energy. Since light energy is the same as electromagnetic energy, it travels through optically transparent conductor, such as glass fiber, an electric insulator due to the extremely high frequency of light energy.

As discussed in earlier parts of this book, optical fiber used for most communications purposes is quite thin and small with a circular cross section. The type of fiber-optic cable mostly used for telephone, television, and data is referred to as single-mode optical fiber. Some early LAN systems used multimode fiber, where the light energy traveled in multipaths along the fiber length.

Fiber-optic communications systems in use today criss-cross the United States and Canada and carry voice, data, and video signals. Many of the television networks connect the central communication points to affiliated stations through leased or contracted fiber-optic systems. There are fiber-optic submarine cables in operation today. Thus countries have an alternative to satellite systems for long-distance communications.

4.3.3.1 Network topology. Fiber-optic communication lines can be used with a variety of network topologies. As for LAN systems today, many wired LANs are still functioning well and have for many years. The change to a fiber-optic LAN has to come from a need for high-speed/high-volume data, voice, and/or video services to warrant the costs. Large companies that have many buildings in a campus or industrial park environment should look at the benefits a fiber-optic LAN can provide. One fiber-optic cable with eight optical fibers could provide a tree-branch topology or a ring-type topology as the case might indicate. Since fiber-optic couplers and signal dividers are available and are completely passive and similar to those used for coaxial cable coupler and divider types, a similar topology can be realized. The cost factors and benefits provided as in any business decision have to be considered.

An illustration of such a fiber-optic network topology is shown in Figure 4.26. This type of ring topology is termed fiber distributed data interface (FDDI) and is described by the standard introduced by the American National Standards Institute (ANSI). This specifies the network as a token-ring type for connecting computing equipment at a data rate of 100 Mbps. The transferring of data takes a lot of both software and hardware to accomplish the task. The software developed for such purposes also has standards, and the need for such standards was addressed earlier by the International Standards Organization (ISO).

The ISO introduced a model for such a system and specified the seven-layer model shown in Figure 4.27. The actual steps in getting data sent would be from the bottom-up. The application layer sets up the task or so-called planned trip, and the presentation layer gets the data to be sent lined up, sort of like dressing for the trip. The session layer addresses such things as getting the data assembled and ready, much like baggage and taxi arrangements for the trip. The transport layer performs end-to-end controls, similar to taking care of problems at the point of departure and arrival of vacation trip. The network layer is important and addresses the routing and timing for the data to be sent, similar to getting to the station for a trip. The data link layer addresses timing and making the data packets, like getting on board and not having everything or everyone. Finally, the physical layer involves the network doing the actual transmission and reception of the data, like the train, bus, or ship for a trip.

Ethernet-type networks also have employed fiber-optic technology to increase their data transfer rates. This type of system conforms to 10 BASE-FP, 10 BASE-FB, or 10 BASE-FL for distances of 1000 to 2000 meters. Presently, Ethernet-type networks can operate at 100 Mbps over fiber cable, much to the

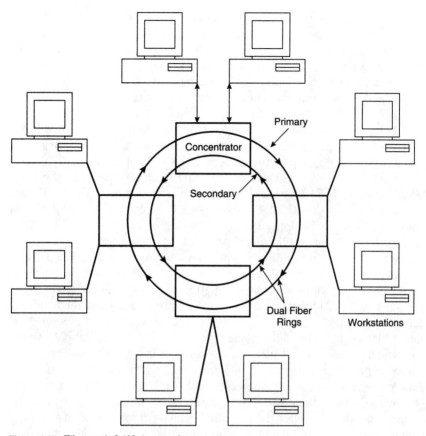

Figure 4.26 Fiber-optic LAN ring topology.

surprise of many. One-hundred Mbps rates for fast Ethernet (10BASE-VG) would also provide this rate over twisted-pair copper cable.

Fiber-optic cable is still cable and essentially acts like a supercable in that huge distances can be realized with no electronics in between, due to low

Layer Number		
1	Physical	Actual Data Transmission Through a Medium Is Made Up of Data Packets; Timing and Error Checking; Routing Control from Data-Source Networks; End-to-End Control; Setting Up Source Nodes; Formatting of Data, Coding, etc.; Special Functions and Purposes Such as Data File Transfer, Teleconferencing, Electronic Mail, etc.
2	Data Link	
3	Network	
4	Transport	
5	Session	
6	Presentation	
7	Application	

Figure 4.27 OSI seven-layer model.

transmission losses. Still, such cable is not perfect. New fiber-optic devices are still being introduced to increase the possibilities of fiber-optic technology. The fiber-optic power divider/coupler enabled fiber-optic cable to be used in a variety of network architectures. The optical power amplifier developed recently is starting to come into use and will make for even larger distances between transmission and receiving points possible.

4.3.3.2 Network control and management. As discussed earlier, network control of fiber-optic cable systems sort of grew out of necessity. The telephone industry developed the Synchronous Optical Network (SONET), which supports a data rate of 10 Gbps (1000 Mbps). However, the basic SONET rate is 51.84 Mbps and is named optical carrier 1 (OC-1). Higher speeds are multiples of OC-1. For example, OC-3 would operate at 3×51.84 Mbps, which is 155.52 Mbps. Specifying the higher rates as multiples of the basic rate maintains synchronism.

The highest rate is at OC-192, which gives a rate close to 10 Gbps. This type of system supports full-motion video, voice, and data in large amounts simply because of the speed of the data transfer. As discussed previously in section 4.1, SONET systems have layers that are necessary in following the general OSI model. Each layer addresses the particular information necessary to implement the data transmission. The SONET digital data frame contains header data that consists of overhead bytes with the layer information. The actual data bytes are arranged in packets transported in the data frames with available empty slots. At the receiving end, the data packets have to be assembled and sent to the correct destination. Each fiber can be used to carry a SONET data channel, which can contain a mix of voice (telephone), video, and data digital information. At this point it should be realized how efficient optical transmission has become.

As mentioned earlier, another type of fiber-optic network topology is the Fiber Distributed Data Interface (FDDI). This network topology is of the token-ring type, and the control method conforms to the OSI model. Header bytes carry the control information. A token-ring type network is very useful for typical LAN systems. However, as a long-haul, high-speed data transportation medium, it just hasn't developed.

Ethernet systems using cable resembles the OSI model. The data layers are divided into the data capsules or packetizing section and the transmit/receive control section. The physical layer is divided into the encoding/decoding section and the transmit/receive channel section. Actually, the model Ethernet more closely follows the IEEE 802.3 Logic Link Control Model.

All LAN systems contain data bits or data bytes that contain identifying information. Such information as source identification, destination identification, length of data words or packets, location of packets in data frames, and timing/synchronizing all are transmitted with the data. At the receiving end, the circuits activated by such header information assemble the data into the specified format for the receiving data system. Long-haul data transmission systems might rearrange, repacketize, and reframe the data several times along the way. The header information traveling with the data prevents loss of data packets or sending data to the incorrect destination. Also, the layer at the physical medium acts like the train conductor looking after boarding passengers and seeing that they are seated.

From the communications engineer's point of view, the high-speed bit stream travelling on the cable or optical fiber should not be deteriorated sufficiently so as to cause bit errors. Errors can cause the header data bytes to be misread, thus causing the reassembling of the data packets to be faulty or data misdirected to incorrect receiving addresses. Network controls are very important to high-speed multiplexed data transmission systems.

4.3.3.3 Optical/electrical interfacing. The end-to-end users of optical cable systems are the usual electronic digital computing systems, which require electrical pulsed signals conforming to the binary zero and one conditions. Such connections might be either in parallel or serial form. The conversion from electrical signals to optical signals has to be done at the transmitting end, and the conversion from optical signals to electrical signals has to be done at the receiving end. The tasks at the transmitting end is the same regardless of the transmission method. The data packets have to be assembled, labelled, and addressed according to the data transmission protocols being followed. For the case of telephone traffic on a SONET system, the DSI levels are assembled or multiplexed to DS3 according to SONET protocol and transmitted optically. At the receiving end, the telephone packets are unlayered, separated out into the DS1 level, and further disassembled into either analog signals to a local exchange or repacketized for another hop. The higher data rates supported by optical permits the higher multiplexing level of the lower DS type of data rates. Fiber-optic methods are a significant contribution to the information highway.

The circuitry involved with optical transmission is quite simple. The optical transmitter is a laser operating at some wavelength as specified by the chemical composition of its elements. The optical power referred to as light power output is not in the region of human visibility. However, it can damage the human eye, so personnel working with such equipment must be protected by special filtering glasses or goggles. The actual solid-state laser is fitted with a lens assembly that will concentrate the radiated energy into an optical fiber. The light output of the laser is varied by the modulating or driving electrical energy containing the data. The glass fiber cable is transporting a single optical carrier with data modulation. If another laser operating at a different wavelength is connected via an optical coupler to the same fiber, then this same fiber can transport another complete optical data channel. This type of optical transmission is known as *wavelength division multiplexing* (WDM) and is illustrated in Figure 4.28.

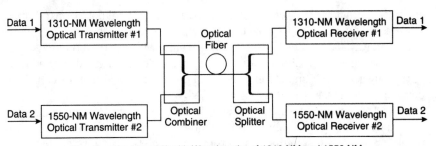

Two Carriers (1 and 2) with Wavelengths of 1310 NM and 1550 NM
Respectively Are Combined and Transmitted through the Optical
Fiber to the Receiving End, Where the Optical Signals Are Divided and
Coupled to Their Respective Receivers.

Figure 4.28 WDM optical transmission.

Until recently, WDM-type transmission was not possible because optical couplers and dividers were not developed. Presently, such couplers and dividers are available with a number of ports. The optical loss per port increases with the number of ports, similar to the case of power dividers and directional couplers used in the coaxial cable RF environment. Higher-power lasers and more sensitive receiving diodes make longer transmission distances possible. The solid-state system manufacturers have definitely risen to the occasion and have made significant contributions to this mode of telecommunications.

5

Developing and Building the Information Highway

Many have said that the information highway is already here. Of course, that is partly true. The present-day telephone network is indeed a bidirectional communications system that supports voice, data, and still or slow-moving video pictures. Some people might regard this system as a sort of "secondary-road-type" information highway. As far as voice traffic and most data traffic is concerned, the telephone network does quite well and is one of the most reliable systems in operation today. But the communication needs are increasing at an alarming rate. Such stop-gap measures as cellular telephones and laptop computers with telephone modems and miniature fax machines have made many an automobile into a portable office. Still, the main carrier is the telephone system. One might wonder just how much the system can take. In order to stay in a competitive market, the telephone industry is definitely busy continually upgrading the system and trying to keep up with the demands. Other communications carriers are also busy upgrading their facilities and promoting their businesses. Thus, goals and specifications need to be addressed to provide direction for the information highway.

5.1 Defining the Information Highway Network

Clearly, the information highway concept should be properly defined as to what the requirements are and what they might be in the future. Since no one has the proverbial crystal-ball, careful projections on the requirements, cost factors, and available technology should be made carefully. The parallel scene between the information highway and the road highway system in the country seems to be a good one. Essentially, we build roads where existing traffic is heavy and the routes are well travelled. The need for expansion of communication facilities should dictate what and when such construction should take place. Clearly, the market forces and the economics of the situation should point to where areas of new telecommunications plants should be built.

5.1.1 Goals/specifications

The problems facing the growth of an information highway are many. The public needs to be educated as to the nature of the concept, along with benefits and cost estimates. An educated public is always a benefit to marketing and promoting an idea. The importance of proper communications facilities should be made known. For example, a major industry or commercial enterprise is going to take a hard look at the available communication facilities before relocating to a community or municipality. One of the reasons large manufacturers and businesses are located in or about major cities is the availability of communications systems. This, of course, might be the result of the "chicken-and-egg" situation. Labor-intensive businesses often locate in city areas due to the labor pool, public transportation, and shippers (either rail or truck) being nearby. Therefore such areas were served with a high level of communications facilities. A highly developed communications highway system will be of benefit to the economic growth of an area and the country in general.

5.1.1.1 Defining the market-economics factors.

Before most companies build or manufacture anything, a marketing analysis is usually performed first. This is a good business practice. Prospective customer surveys are often a useful tool used by marketers of a product. The information highway is really no different. Common carriers should analyze various areas of their systems for areas that need to be fixed or upgraded. So much of an information highway is already in place and operational. Tying into and connecting is the present problem. For the parallel to our road structure, some on-ramps to the information highway are like dirt roads, e.g., no easy access. Clearly, such construction has to be done to correct such communication bottlenecks.

Since, for the most part, free enterprise exists in the country today, communication facilities consist of many different companies and systems. For a telephone call to take place across the country, most likely several different companies will be involved. Telephone system compatibility from system to system was the result of mainly the Bell System standards. No problem with any cross-country connections resulted.

The Bell System was a guiding force in the formation of the nation's communications policy. Standards for equipment performance, system performance, and personnel performance were set by the Bell Systems. An extremely highly reliable standard for message traffic (99.9%) was set and maintained. When communications were needed, the Bell System was there. With the breakup, which essentially defused the companies into many separate entities, the existence of such tremendous influence on the communications area has diminished. The U.S. government has taken some interest in the communications highway through Vice President Al Gore's office. Both houses of Congress have committees on telecommunications. Still the industry is looking for some kind of leadership. At present, market forces seem to be the controlling element.

Manufacturers of telecommunications equipment are also a definite force. They are trying to second-guess the equipment needs of the communication system operators. Often the manufacturers make what they think the operators need, and then the manufacturers go out and try and sell the equipment. Both producers of software and the microchip manufacturers are also trying to second-guess the market. In short, there are a lot of spinning wheels trying to go somewhere. However, the good news is that some concepts have shaken out and look very promising in assuming a place in the information highway.

5.1.1.2 Public relations/customer education. Since there is no one main guiding force to address the information highway, the communications operating companies should take control. First a good public-relations plan should be put in place, possibly through a telecommunications society or association. The general public should be made aware of the concept, the plan, and the rate structure. Most families do not want to bear horrendous charges for their children to play video/computer games with friends across the country. Commercial businesses will want information on types of services, computer data, voice (telephone), and/or video services with proposed rates and charges. This same public relations activity can also, through prospective customer dialogue, provide information about communication needs and desires. This is really a two-way street.

Since there is no megacommunication company to lead the way, it is hoped that the government will act as a guiding force. Unfortunately, regulating legislation is being proposed, and the lobbying forces are hard at work. The common carriers and the cable television operators are all maneuvering to get into a top spot on the information highway. Subsequent sections of this book will address the connectivity problems between the various communication methods.

5.1.1.3 Customer equipment and system interfacing. The customers of communication systems consist of both business and domestic users. Presently computers are connected to a communications system through a computer modem that is either internal or in a separate unit. This modem is in most cases connected to the local exchange carrier (LEC), which is the telephone system. Some commercial users have access through their own LAN system to fiber-optic facilities often provided through the LEC. A device that stacks and layers the computer data for transmission over the fiber-optic cable is required as an interface. There are many types of interface devices manufactured today. These devices need internal microprocessors programmed to accept data from LANs using Ethernet or other types of LAN system data. The data has to be sent serially at high speed over most physical communication systems. This should be evident because the data is being transmitted over one fiber.

At the receiving end, the receiving network interface unit (NIU) or modem converts the optical signal to an electrical signal and then selects the proper data words and prepares them for either serial or parallel data to the proper work terminal. As the information highway develops, there will undoubtedly be many new interface devices that need to be invented to provide the higher speeds and lower error rates. This results in more business for equipment manufacturers and software producers.

5.1.2 Present system interconnecting

The first task is to use the present systems in place today to do the job required. This has probably been going on since before the information highway came into vogue. Use of satellite communications has met many commercial needs. Such facilities might be regarded as an airport in the information highway. Similar to normal travel, it might be more expensive to fly, but it might be the only choice. Also, personal communications and/or cellular telephone is now popular and commonly used. It used to be a status symbol to have a car phone or pocket phone. But now the price has dropped to the point they are within the financial reach of most everyone. This is, again, a bridge into the telephone system.

Connecting present systems into a general-purpose, high-speed telecommunication system is what is known as the information highway, often referred to as the Infobahn, or more elegantly called national information infrastructure (NII). Information interchange is being carried out by many involved with data or computer information transfer. Internet is probably the most well known. To make it easier for most home and business computers to work with Internet, some excellent software has been developed that makes browsing and perusal a lot easier. Commercial users of Internet provide financial services to Internet participants. Credit information is carried encrypted so only the participants with keys can decode the data. The Internet system is loaded with information about products, goods, and services that are for sale. Often, ordering and billing can be done as well.

Presently the Internet is operating on existing communication facilities, but for the domestic home user and many small businesses, the gateway is through the local telephone exchange. There are, in general, three types of cables coming into homes and many commercial buildings: electric wire, cable television wire, and telephone wire. The only cable with any decent amount of bandwidth is the cable television wire. However, only limited upstream bandwidth is available. The cable system, for the most part, is a one-way delivery system for television programming.

It would seem appropriate that the telephone company would replace the telephone wire with a combination coaxial twisted-pair duplex wire. This would enable high-speed up/down digital data to be carried on the coaxial cable section, and ordinary telephone communications would go through the twisted-pair section. Naturally, the two types of plant have to be carried throughout the pole line. How much cable operators get involved with duplex data transmission remains to be seen. For the most part, cable television operators seem most interested in video on demand (VOD) and/or some form of interactive television service. As soon as the regulatory climate shakes out, it will become more clear who is going to provide what to the communications marketplace. Several cable television operators are expecting in the future to enter into the telephone business as soon as proposed telecommunications legislation is passed through government channels permitting such use. Plans have been made technically to expand the cable television system return or upstream path required for full two-way (duplex) telephone service. Cable operators also will be using this return path to enhance their video on demand (VOD) and/or interactive television service.

5.1.2.1 Bridging and network interconnects. It is no simple task to connect the pieces of the various cable communications systems together into an information highway. Not only is one faced with the technical differences with the several types of systems, but also with the owners or corporate entities that control these systems. The laws governing telecommunication carriers are also an extremely important factor in forming the interconnections. Work has to be done in many areas to get plans and agreements ready and in place.

The cable television industry is definitely positioning itself to be a major player in the information highway. Several large multiple system operators (MSO) have been involved with mergers and buy-outs with other companies. It seems inevitable that only a few very large systems will remain active in the cable television business.

The telephone industry is most likely in the strongest position because essentially it is in the business of the information highway with telephone and computer data traffic already travelling its lines and cables. However, from a bandwidth point of view, the lines of the local exchange are not wide-band and hence will not support high-speed digital data or any signals requiring large amounts of bandwidth. Present telecommunication laws prevent the telephone industry from using profits from such service to support other services such as video and high-speed data. Such profits are spent in improving telephone services or lowering telephone costs to consumers. If legislation is passed permitting such spending, then new services will be developed by the telephone industry.

High-speed digital data is presently transmitted by privately owned LAN systems using one or more of the developed protocols such as Ethernet, Fast-Ethernet, PC LAN, MAP/TOP, etc. The telephone industry has been actively using SONET technology to multiplex digitized voice channels over fiber-optic cable. As previously discussed, this SONET system is a synchronous method of transmission that takes asynchronous data channels made from the DS-1 standard through DS-3 and packetizes them onto a SONET system.

Another type of system that allows asynchronous data to be transmitted is referred to in the industry as asynchronous transfer mode (ATM). Such a method is similar to the telephone exchange switching technique and is often referred to as the ATM switch. The main difference is that the ATM switch is a smart switch in that it is self-routing. Again, the chip manufacturers are responding with a variety of ATM switching devices. The operating speed of available devices supports data rates of up to 400 Mbps. This is enough to transmit full-motion video and data. The ATM technology started as a LAN switching technology where the routes and bandwidth are ordered and set by the switch controller. Still, the ATM system is in its infancy, and some industry gurus even today are forecasting its failure.

Other technologies known as Fiber Distributed Data Interface (FDDI) are considered as an opposing technology. As previously discussed, FDDI is a dual-ring type of topology with token passing. In order to keep the so-called air clear, a more detailed look at the various topologies and protocols are in order.

As mentioned before, on several occasions the telephone industry became involved with digital data transmission. Telephone subscribers could communicate by use of the computer modem. The telephone industry also was using digital transmission methods referred to as Integrated Digital Network (IDN) for telephone traffic, simply because digital signal transmission was integrated over the telephone system. The data rate of 64 Kbps was implemented over normal twisted-pair telephone lines. This IDN concept was expanded to the Integrated Systems Digital Network (ISDN), where voice and data were converted to digital coded signals and transmitted over wired networks. Since the wired networks were mostly telephone-type wire, data transmission was slow. As the need for higher data rates increased, the ISDN system was expanded to the Broadband ISDN system

(B-ISDN). Higher data rates were now possible and great enough to carry multiple voice channels, data, and video.

The B-ISDN implementation is in CCITT recommendation I.121, which distinguishes between interactive or distribution-type services. Interactive services fall into three categories: conversational, message, and retrieval services. Conversational services are two-way services that include interchange of video, text, and data message services, including message handling, mailbox (e-mail) services. Retrieval service includes databank, videotex electronic advertising, and video library services. Distribution services are often referred to as a broadcast-type service where user involvement might not be required. Services such as cable television and electronic newspaper service do not require user presentation control. Services such as teletex and/or video text would require user control.

The transmission methods to implement B-ISDN services can use what is known as fiber to the home (FTTH) or synchronous digital hierarchy (SDH) or asynchronous transfer mode (ATM). FTTH is an optical fiber transmission method that requires the expansion of fiber-optic systems to the home or office. Such systems can be of any type of network topology such as ring or star. SDH uses SONET methods that will act as long-distance digital trunking of voice/data and video services. The ATM technique will set up the transmission path and assemble the asynchronous data into packets to form cells. ATM is a data packet switching method that shows much promise as an information highway player. The ATM technology supports SDH/SONET and the B-ISDN implementation.

System interconnecting for digital transmissions from one network type to the other is necessary to complete the information highway from end to end. Where some interfacing from older, established LAN systems is required, some type of bridge devices will need to be installed. Such bridging between the electronic, wired systems and fiber-optic systems in many cases has been developed and is in operation today. Figure 5.1 shows areas of connection at various points in a system.

5.1.2.2 Plant additions and system upgrades. The major transmission facilities in place today are the long-distance fiber-optic systems spanning the country. These systems presently carry high-speed digital signals with voice, data, and video services. Therefore the major superhighways of the information highway are in place. These are toll roads with rates and charges set. Access to these facilities are through local and secondary systems to the so-called on-ramps. In many cases, there is work needed in the local exchange area to expand the necessary bandwidth for high-speed digital data. Some communication facilities, such as fiber-optic cable owned by the commercial power industry, can provide important paths needed to connect local public exchanges to existing long-distance fiber-optic systems. Unfortunately, the local telephone companies cannot use profits from

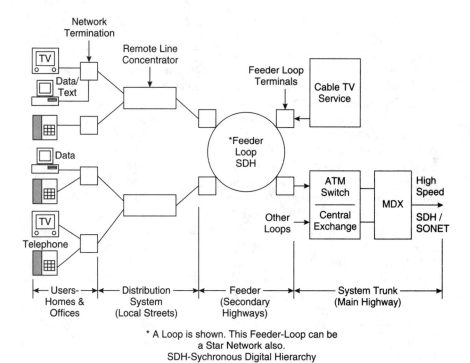

Figure 5.1 User to high-speed network.

telephone services to finance system upgrades that do not directly affect telephone service rates and charges. If and when congressional deregulation permits such expenditures, plant upgrades to the so-called fiber to the home (FTTH) status will be slow in coming.

Cable television operators will have to increase the return path bandwidth of their cable systems to support two-way, full-duplex operation. Most fiber-optic systems installed by the cable television industry are amplitude modulated wideband analog systems. This means that the whole cable television frequency band containing the RF on-channel television carriers intensity modulate the laser optical transmitters.

The receiver is often a utility-pole-mounted device. At the receiver, the RF carriers containing the television programming are recovered intact and connected into the broadband cable television amplifier cascade. This is an efficient method because there is no departure from the standard television channels that a subscriber television set is able to be tuned. Extra fibers available in this same cable structure can be used as a return-path conduit to the cable system head-end and ultimately to the local telephone exchange carrier. Some of the larger and more technically progressive cable operators have sufficient fiber-optic cable in place to support a variety of communication facilities. Microcellular telephone service is one use of the expanded upstream cable television system.

5.1.2.3 Network integration and control. Network integration and control for the formation of an information highway is an important factor in putting it all together. Since the presently installed and operating communication facilities in the country are owned and controlled, under the present laws, by many different companies, coordination of the efforts and areas of responsibilities is a large problem. Not only are there many technical problems in joining the various LANs, telephone service networks, and cable systems, but there are also problems in working with the many corporate entities. Clearly some form of leadership would help. However, it seems that market forces alone are presently the only factors determining how the information highway will develop. Some studies have indicated that only one broadband network as a sole provider of voice, video, and data services to the public will be able to survive economically. In short, competition among various providers and methods will cause duplication of efforts, and the resulting unnecessary costs will ultimately be borne by the consumer.

Some bridging device is necessary for connecting systems such as Ethernet and/or token-ring-type LAN systems to, for example, a SONET high-speed transportation system. What the input to a bridging device must look like depends on the network, which acts as a source node. The output will have to match the transportation system. Some of this equipment has been developed and some has not. Many types of local area networks are in operation today. Some are most likely to have been in operation a long time and hence operate at low data rates. Connection to an information highway operating at very high data rates will involve a bridging device with large amounts of data storage needed to accumulate the information from the low-data-rate LAN and packetize the data according to the required high-speed format. Such equipment is presently appearing on the market. Again, the industry is reading the correct signs and responding with products to meet the needs for the interconnection between user and data transmission system. Integration is essentially being supplied on an "as-needed" basis. Carriers of high-speed voice, video, and data transmissions are carried out by several telecommunications companies such as AT&T, MCI, Sprint, and Wiltel Communications. As time goes on, more will be added.

Satellite communications are still very important in carrying long-distance communication, and satellites are the only carriers where no cable facilities, either land or undersea, exist. Digital transmission is being used more and more because digital signals are more robust in a noisy environment. In coaxial cable systems used by the telephone industry, the former double sideband suppressed carrier (DSSC) methods are being phased out by multiples of carriers carrying quadrature amplitude modulation (QAM) techniques. Some cable television operators have found that compressed digital video signals such as MPEG II can be carried on a cable system using QAM-64 alongside standard television carriers using the regular vestigial sideband AM modulation. High-definition television is proposed to be a digital signal, compressed and transmitted most likely as a QAM-64-type method of modulation. Some of these television signals will be distributed by satellite methods in scrambled (encrypted form) to cable television systems for distribution to subscribers. AT&T has invested heavily in satellite systems that carry much voice, data, and video information in digital form. Integration of the many forms of communication signals and systems is a major portion of the formation of the information highway.

5.2 Construction, Maintenance, and Training Concerns

System components, maintenance and training concerns are important areas for the information highway. For the most part, new components and systems are being developed on an as-needed basis. However, some manufacturers of components offer products that are needed to bridge from one type of system to another. This supplies a ready solution for many LAN interconnect problems. Test equipment manufacturers have supplied much equipment for the telecommunications industry in the area of wire, cable, and fiber-optic systems. Availability of components and equipment has been discussed in many magazines and technical journals. Technical articles and tutorial papers have helped to educate and inform the technical community of progress in the telecommunications industry. Most technical people and many people in the telecommunications business world read many of the periodicals and magazines covering the field. This up-to-the-minute, state-of-the-art status on the direction the telecommunications industry is heading provides important information to the many people and companies involved.

Professional societies such as the Institute of Electrical Electronic Engineers (IEEE), with its many professional groups, have been and are very influential in the development of the telecommunications field. The IEEE has been instrumental in setting various standards for the telecommunications industry, and these standards have prevented total confusion. Known as the 800 series, these standards provide the rules and protocols for interconnection of LAN systems. Thus compatibility was achieved. The cable television industry founded the Society of Cable Television Engineers (SCTE), which has been in existence for over 20 years. The SCTE has established a training program and a technical certification program and is working on standards through the American National Standards Institute (ANSI). One must not forget the area of government regulations through the Federal Communications Commission (FCC) and the telecommunications laws. On the other hand, the federal government has done little in the form of either grants and/or tax relief or incentives to foster the growth of the information highway. Each one of these areas of concern will be addressed in the following sections.

Last but not least is the question of where are all the technical people going to come from. The telecommunications engineers and technicians are going to need training and expertise in more than one area. The telephone industry is probably in better condition than most companies because they have and are involved in such a variety of communication methods. However, at the local exchange level, technical personnel know little or nothing about fiber-optic systems or cable television broadband coaxial cable systems. If fiber to the curb (FTTC) systems are built, local line technicians will have to become properly trained in fiber-optic system maintenance. Most technical training programs are company in-house training programs consisting of classroom time and on-the-job training (OJT). Technical schools should be encouraged by industry and government to offer students technical training at all levels.

5.2.1 Coordination of communication corners

The many companies and enterprises involved with telecommunications fall into three categories: the system operators, the equipment manufacturers, and the software companies. The parts played by these companies depend primarily on market forces. The demand for telecommunications services from the system operators causes equipment and software orders and purchases from the suppliers. Advertising and marketing efforts have paid off in making the public aware of the various telecommunications services offered to commercial and industrial enterprises.

The many professional societies, both technical and business, have taken an active role in educating and informing its members on the merits and accomplishments of the telecommunications industry. In general, the professional societies have been a guiding force in the development of the country's telecommunications systems. Most professional people should be members of a professional society, the benefits of which are the journals and publications and the various meetings and seminars. Keeping up with the progress of one's profession through continuing education is most necessary in attaining a successful vocation.

The federal government has been active in telecommunications through the rules and regulations of the Federal Communications Commission (FCC). This commission was formed with the passage of the Communications Act of 1934. Since then, laws have been passed broadening the scope and areas of authority of the commission (FCC). Both houses of Congress have committees on telecommunications that advise members of Congress on communication matters. Action by the FCC is supposed to be for the "good of the public" or "in the interest of the public."

5.2.1.1 Telecommunications companies.   Private industry operating in the telecommunications area has been and is a major force in building an information highway. The telecommunications system operators have con-

structed many types of networks using a variety of topologies, methods of signal transmission, and types of wires and cables. The satellite telecommunications facilities have grown to large proportions and carry voice, video, and data. Also, satellite systems have been used and are presently used to provide navigational and global positioning information and photographic information for the earth's surface. This photographic information is useful for defense purposes and environmental studies of the polar ice cap, the ozone layer, and weather warning services, to name a few.

The telephone companies and the cable television operators make up the bulk of the wired system. The telephone industry has been investing heavily in fiber-optic cables. Also, many cable television operators are using fiber-optic technology in several areas of system application. When one considers the signal bandwidth of commercially available fiber-optic cable and the signal bandwidth of the radio frequency spectrum, it is clear that more bandwidth exists in fiber-optic cable. Over-the-air transmission should be reserved for applications where they are most effective, such as mobile transportation communications; satellite navigation, positioning, and surveillance service; and space travel and research. Theoretically, most of the communication needs of the world can be serviced by the mix of fiber and metallic cable systems. This will be, for the most part, the backbone of the information highway. Off-air radio transmission can be reserved for communications to areas where little or no cables exist, such as third-world countries and remote areas. Short-range radio transmission methods such as cellular telephone and personal paging will act as bridges and feeds into the cable systems. The over-the-air radio spectrum should be regarded as an important global resource, and its uses should be reserved for applications it serves best.

The problem at hand now is to interconnect the cabled systems. This is necessary to provide access to the high-speed digital information transportation routes forming the telecommunications superhighways. The hurdles are resolving the interconnection of various private networks operating with a variety of protocols and data packeting methods. In many cases, bridging devices are presently available that perform these functions. The television industry has accepted the MPEG-2 standard for digital video compression, and the chip sets are available that perform the compression, decompression functions.

Video information using MPEG-2 digital compressed format can be transmitted using quadrature amplitude modulation (QAM) techniques or vestigial sideband modulation. These systems are called 64QAM and 16VSB, types of signal modulation. A short tutorial on QAM methods is given in appendix N. The 64QAM method allows for compressed digital data of more than 20 video channels to be carried digitally in one 6-MHz standard television cable channel. Tests indicate that the expanded electrical power necessary for the eight amplitude levels for 64QAM results in the ability of normal, standard cable television channels to be transmitted on the same cable system alongside the 64QAM channels. There are still conflicting camps concerning the merits of the 64QAM and 16VSB method of video transmission. Through the fiber-optic SONET system, the telephone systems are able to carry packets of digital signals containing MPEG-2 compressed video. Now the trick is to assemble the high-speed packets and transfer this data to, for example, a cable system head-end for 64QAM or 16VSB transmission to subscribers. The set-top terminal now has the task of converting the digital data to NTSC analog signals required by the present television sets.

HDTV operating with an aspect ratio of 9:16 will probably be digital, and subscribers' HDTV sets will probably be digital. The digital signals will contain information on the position, brightness, and color of tiny pixels making up the television screen. A comparison of the equivalent screen viewing area for the 3:4 and the 9:16 screens is shown in Figure 5.2. Notice that as the area increases, the width of the screen increases more than the height increases for the 16:9 aspect ratio. This will use much more wall space than conventional 3:4 aspect-ratio television sets. This problem does not affect the transmission method, and digitally compressed 3:4 aspect-ratio pictures could be carried on present cabled systems.

The most common problem facing the information highway is the issue of equal upstream and downstream signal bandwidth and the resulting data rates. The first question: is equal upstream and downstream bandwidth necessary? Most upstream traffic from domestic homes is now handled at available computer modem rates at maximum of 28.8 Kbps. Commercial-industrial data from area LAN systems could require a broadband upstream path. Commercial areas, industrial parks, and college campuses most likely will require a large upstream-downstream transmission capacity. The local exchange carriers (LECs) usually have and will provide bidirectional communications facilities to such areas. Thus the connection for high-speed, bidirectional data access to the information highway for many business ar-

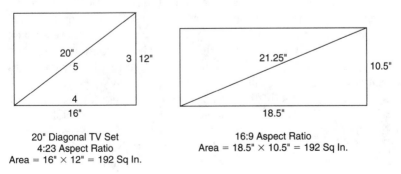

Figure 5.2 Television screen size for 4:3 and 16:9 aspect ratios.

eas is already provided. Areas of industrial development will need to make sure that proper telecommunication facilities will be in place for companies relocating to the area.

5.2.1.2 Professional society involvements. Professional society involvement in the information highway has been there from the beginning. One of the largest professional societies is the IEEE, which was founded by the joining of the Institute of Radio Engineers (IRE) and the American Institute of Electrical Engineers. Early on, the IEEE initiated standards for methods of computer interconnections and standards relating to many areas of telecommunications, including broadcasting and satellite communications. This society has developed a series of professional groups or societies within the parent society. Each one of these professional groups has a specific area of expertise, such as the Power Engineering Society, the Communications Society, etc. The IEEE has over 300,000 members throughout the world and is more of a world society, although it is headquartered in New Jersey.

The Society of Motion Picture and Television Engineers (SMPTE) has been and is very influential in video and television engineering and standards. This society started with the technical people involved with the movie industry. With the invention of television, the society later included the television broadcasting industry. With the development of video graphics, both the broadcasters and the motion picture people use graphics software to generate cartoon characters and much of the animation seen in the movies and in television. Many of the network television stations convert all the analog signals to digital signals and switch and edit the signals in the digital domain before converting them back to the analog domain for broadcast transmission. If a truly digital television monitor or receiver becomes commercially available, then digital transmission could result.

The society of cable television engineers is a society made up of technical people working or associated with the cable television industry. This society has been in existence since the early '70s and has some 13,000 members. Serving its members has been one of this societies strong points with its many seminars, expos and meetings. Local chapters, often meeting monthly, offer timely technical programs and tutorial lectures. Since many technical schools do not provide many training programs to the industry, there is a lack of diplomas or certificates demonstrating a technician's expertise. Therefore this society has developed a certification program that certifies its members as to their level of accomplishment. This is acquired by examination only. The SCTE is becoming active in the proposal and adoption of its industry standards through the American National Standards Institute (ANSI).

Another professional society involved with video and audio signals is the Society of Broadcast Engineers (SBE). This society is active in technical standards and testing and measurement methods pertaining to the broadcasting industry. Signal processing, switching, and control methods for both analog and digital signals are the so-called hot topics in the broadcast industry. This society also has local chapters, meetings, seminars, and publications serving its members.

The National Association of Broadcasters (NAB) is a professional society that has been around for a long time. The NAB has been active in setting broadcasting standards and methods since the early days of radio. The NAB handbook contains a wealth of technical information and is upgraded every so often by the issue of a new edition. Most broadcasting engineers have a copy of one the editions of this handbook. The NAB also conducts area meetings periodically, with the national meeting being the largest and most popular.

There are many smaller societies that serve the telecommunications industry, such as the early telephone-oriented organizations. The newest societies and organizations are those pertaining to computing, both hardware and software. Periodic meetings, seminars, and short tutorial courses sponsored by the many professional societies and groups have been instrumental in providing continuing and ongoing education to the industry.

5.2.1.3 Government controls and regulations. Here in the United States, the federal government has to be active in regulating and controlling the telecommunications industry. Historically, the start was the communications act of 1934, which authorized the Federal Communications Commission (FCC). The telephone/telegraph industry, it was felt, needed some controls to protect the public interest. Also, the radio broadcast industry, which was growing quickly, needed more regulations pertaining to radiated power, frequency accuracy, and in general the area of transmission. The commission established standards for classes of broadcast stations. Designated stations of a certain class were allowed a selection of frequencies, transmitted powers, antenna gains, and radiational areas. The area of broadcast coverage was and is divided into market areas. Broadcast stations were permitted to cover these specific areas (and no more) with their radiated signals. As radio propagation for standard AM radio changed from day to nighttime, stations were designated a daytime power output and a nighttime power output. This is an example of the areas of concern for the FCC in the early days of telephone and radio.

Today the FCC rules and regulations cover all facets of telecommunications and are contained in volumes. Conversion of these rules and regulations are or have been reduced to CD-ROM libraries. Many consultants and attorneys that represent clients before the commission have access to all the rules and regulations. In most cases, in this day of specialization, these consultants and attorneys specialize their practice in the many areas of telecommunications. Although the general public can petition the commission, it is often more expeditious to use a consultant or attorney to act for a station application for license.

As mentioned before, Congress has telecommunications subcommittees to advise and/or assist the members of Congress. Many of the laws governing the industry started in one of these committees. Financially, the industry has from time to time been allowed assistance through tax relief, rules governing return on investment, and sponsored research and/or development grants.

Probably the most significant legislation in the telecommunications industry is the breakup and restructuring of the American Telephone and Telegraph (AT&T) company. The seeds of this happening were sown by the earlier Carter phone court decision that allowed other types of products and equipment to be connected to the telephone network. The net result of this action allowed for competition in the telephone business, both in service and equipment. AT&T, on the other hand, was permitted to enter areas of business that were previously prohibited. Gone is the influential, single megacorporation that was so effective in developing world standards for communications and equipment. However, what has emerged are highly effective and efficient separate enterprises that can and are competing for a large share of the telecommunications market. Books and magazine articles have been written on the Bell Company breakup, and there are pro and con positions. However, the consortium of the companies such as the regional Bell operating companies (RBOCs), the Bellcore Company (Bell Communications Research), which operates the research (Bell Labs) and the manufacturing division (Western Electric) are still active participants in the telecommunications business. Now equipment can be sold and marketed, and this was previously prohibited. Still at issue is the long-distance telephone business, which is contested by other telephone operating companies. It is expected that congressional legislation will be forthcoming, and many issues will be settled at that time.

The cable television industry has also been the subject of federal legislation. Some of the largest cable operators with systems in heavily populated and industrialized areas are trying to enter the telephone industry by combining telephone signals on their cable system. By the same token, the telephone companies are attempting to provide video services such as movies, games, catalog, and/or library type services by expanding their local exchange area bandwidth. Also, the FCC has increased the number of testing procedures and the schedule of performance testing for cable operators. System leakage of signals is controlled by a continual measurement, testing, and repair program for all cable operators. Records and files must be kept to prove compliance with the leakage control program, and heavy fines and penalties can be enforced on the violating operators. The FCC also is involved in the rates and charges cable operators impose on their subscribers. Recent rulings have caused many cable operators to lower their rates for some services.

The federal government is active in the development of the telecommunications system in this country. Vice President Al Gore has been visible championing the information highway. Still the market forces are very strong, and just what people will buy and how much they are willing to pay for it is still very unknown.

Some network operators such as Internet American, On Line, and CompuServe offer subscriber services such as e-mail, information bulletin boards, and catalog shopping, to name a few. This type of service is often referred to as the information highway, and at this point in time this might be true. However, as the network is put together and bottlenecks disappear then high-speed data and video services will appear.

5.2.2 Personnel education and training

In the past the telephone operator's maintenance and construction crews kept the network system operational. However, with so many divisions and sections of the operations of the telephone system, many different personnel crews with specific training and skills were needed to maintain the many facilities. At the local level, the line technicians performed the service procedures. The cable television operators also had their local technicians, who were trained in broadband coaxial cable techniques, perform the repairs and service calls in the local area. If a one-network system evolves, the question arises: who is best trained to perform maintenance and servicing? Clearly some cross training seems necessary, and some steps have been taken to do it.

The telephone company has always been a strong advocate for education. In the past, they have gone after the best college graduates, and when hired, the phone company provided more training either on the job or for graduate programs offered by local colleges and universities. Bell Labs has also operated advance training programs for AT&T employers and now offers such training to other company's technical personnel for a nominal charge. Through continuing education programs, telephone company employees were able to advance through the company ranks.

Unfortunately, the cable television industry has not been so fortunate, and this is most likely attributable to its more humble beginnings. Without taking up much time discussing the growing pains of an industry, it is important to identify where it is today. The cable television industry today is made up of a few, very large, multiple system operators (MSOs) who provide cable television service to the large metropolitan and suburban areas serving the greater portion of the country's subscribers. There still are many medium to small cable television operators in business today, and due to buy-outs and mergers, this number becomes smaller and smaller. Many cable television systems got their start as local entrepreneurs who had the political position needed to get the local license or franchise. In many cases, the company was a proprietorship or some form of partnership. Others formed various types of corporations.

Financing was always a problem, and merging of the smaller companies became a solution. In many cases professional management and business skills were lacking, causing many of the fledgling systems to merge. Technical personnel was and still is a problem. Many of the small cable operators contracted out system construction and subscriber installation work to independent contractors, many of which were in and out of the job with the money very quickly. There have been many situations where cable operators were left with little or no technical help when the contractors were finished. In many cases the work was done poorly and in a hurry. Some cable operators hired former telephone company line technicians, former U.S. Army Signal Corp personnel, and former electric company line personnel. This type of person was experienced in working on utility pole lines. As far as cable television head-end and line technicians were concerned, very few professional training programs were available. Technical training is still a problem in the cable television industry, and as the systems get more and more complicated, often requiring sophisticated test equipment and procedures, many systems will have problems in the technical area. Three possible solutions are a better on-the-job training program; an increased number of technical school courses; and proper performance standards, tests, and measurements.

5.2.2.1 On-the-job/in-house training. On-the-job training (OJT) was and probably still is one of the main sources of training used by the cable television industry. The effectiveness of OJT depends on the expertise of the personnel doing the training. Systems with high-quality senior technicians with proper teaching expertise will naturally produce a better grade of technical personnel. If the technicians require more or remedial training, the cable company should provide the training to the level required by the company technical officer. In this manner, cable operators essentially hired and trained their own technical staff.

Many smaller cable operators would try and hire away another company's technical personnel, often referred to as "piracy of personnel." Therefore many cable operators became reluctant to send their technicians to meetings, seminars, and conventions for fear of losing key employees. Thus a bad situation became worse.

From the very beginning, the telephone company provided a form of OJT. In-house company training programs on an ongoing basis provided training for the technical personnel. New employees were required to take introductory and orientation courses on company policy and procedures. Then technical courses for the required equipment would provide the necessary technical knowledge. As new equipment was introduced, in-house training courses provided the retraining for the technical staff responsible for equipment performance. The telephone company maintained a large technical staff with expertise in a variety of areas, which they could draw from to provide technical teachers for the in-house training programs. Since quality was always at the head of the list, the telephone company training program was excellent and provided many high-level technical people for company operations. Now many technical programs offered by the Bell System are also offered to nonemployees for nominal charges. This type of cross training allows personnel in other areas of the telecommunications industry to participate in learning the telephone industry methods and techniques.

5.2.2.2 Private/technical school programs. Technical education in the telecommunications field is unfortunately lacking in some areas. At most colleges and universities, emphasis has shifted from electrical engineering to what is called electrical and computer engineering. In most cases the term computer engineering is used somewhere in the name of the course. At least this fact means that students are going to become heavily indoctrinated in the digital world and familiar with LAN and digital communications. However, the communications theory and technology pertaining to signal carriers and modulation/demodulation technology has to be included somewhere in the curriculum to properly prepare students in the field of telecommunications. Colleges and universities have to conform to the accrediting agencies' requirements for course content in order to maintain the accredited status. However, there is some latitude in the elective courses offered, so students will have the choice of some amount of specialization. Telecommunications subjects certainly should be offered in this area. Colleges and universities should maintain a communication channel to industry to be able to train students for local employment. Listening to the needs of industry is important in the formation of the subjects to be taught at local colleges, universities, and technical schools.

Most trade and vocational high schools offer courses in computer maintenance and repair. Graduates of these types of schools work maintaining computer equipment, faxes, and/or copying machines. In many instances the trade and vocational schools do not offer course work in the consumer electronics field. This can cause a shortage of radio and television technicians. In some states, a license is required to work on radio and television equipment, which raises the question of where training is available. Again, the trade and vocational schools should listen to industry's needs and respond accordingly.

The telecommunications field needs a properly educated technical staff in order to survive in a competitive climate. Today the telephone industry, thanks to the Bell System, has the greatest number of technical experts. This clearly should be a technical edge. The telephone industry has spent enormous amounts of money on education, both in-house training and through participating colleges, universities, and technical schools.

The cable television technical people often were drawn from the radio and television industry, the local electric utility, or the telephone industry.

In most cases the technical staff of the cable television industry had little or no formal electrical/electronic training but had the ability to climb up and down the utility poles safely. As discussed earlier, some cable operators did provide in-house training or tuition expenses for correspondence schools or home-study programs. The National Cable Television Institute (NCTI) offers correspondence courses starting at the installer/technician level and progressing to the chief technician level. This is an excellent program, and when supported by a proper level of "on-the-job" training, produces many competent technicians.

Still, the cable television industry has insufficient numbers of college or university-trained technical people. In the past this was probably not necessary. However, as the telecommunications industry progresses and gets more complicated, more highly trained people will be required. The cable television industry has been fortunate to be supported by a responsive manufacturing force. In many instances the equipment manufacturers spent time and money trying to figure out the equipment that would be required as the cable systems developed. Essentially they were there with the solution before the operators knew it was needed. The manufacturers of cable equipment did much in the form of providing educational material and papers and often conducted free seminars and/or provided educational programs at trade expos and conventions. Several of the large equipment manufacturers offered installation of their products in the form of installation assistance. Some of these corporations offered what was known as turnkey packages where the manufacturer provided all materials and contractor or in-house labor. They built the system, tested it, and passed it over to the owner. In the meantime, the owner hired and trained installer/technicians, often with the help of the manufacturer building the system. Many cable television systems were built in this manner. A turnkey contractor was in most cases the most expensive but best way to go. In many cases the banks or lending institutions, being wary of the fledgling business, required a turnkey contractor. The contractor's staff often provided training to the system owner's technical personnel.

5.2.2.3 Professional society educational programs. The professional societies serving the telecommunications industry have and will always provide significant ongoing training in the form of technical magazines and publications and technical papers presented at trade expositions, conventions, seminars, and meetings. It is indeed important for the industry's technical people to be members of a professional society. Professional participation in the form of technical papers and presentations at meetings gives members of the technical community an opportunity to become known as experts in the field. The old cliche "publish or perish" still has a lot of meaning in today's competitive business climate.

The IEEE is indeed the largest professional society covering the telecommunications industry. Its membership requires proper education in engineering or mathematics/physics. Often members have achieved one or more master's degrees, and many have a doctorate. Again, the many companies and businesses operating in the telecommunications area have been very supportive of the IEEE. This is a case of shaking the hand that feeds you. The IEEE has 11 technical societies within the central organization. The Communications Society has its own separate publication *Communications* magazine. Articles and papers covering the various areas of communications are found in this magazine. The IEEE publishes a general magazine, *Spectrum*, which, true to its name, covers major advances and news covering the spectrum of the electronics field. A separate newspaper named *The Institute* covers topics pertaining to the operation of the IEEE. The IEEE also operates the IEEE Press, which aids its members in the publication of books as joint ventures. Many books by many authors have been published by the IEEE press and are marketed and sold by advertisements placed in the society's magazines.

The Society of Cable Television Engineers (SCTE) is a smaller society serving the cable television industry's technical people. This society is close to 25 years old and has some 13,000 members. The official trade journal is *Communications Technology*, a CT publication that is a monthly magazine containing technical articles and new product information. The society's newsletter carries society information and is also published monthly. The SCTE has available a series of books and tapes covering technical topics at a nominal cost to members. Although the SCTE has the word engineers in its name, few of its members actually hold engineering degrees. Most are technicians at various levels of expertise, but they generally work in an engineering environment. The SCTE is attempting to align the industry's technical community to some standards of technical achievement and has developed a certification program. This program starts at the installer level and progresses through to the engineer level. The many area chapters of the society offer technical programs and tutorial sessions that prepare members for the examinations. A member can proceed to a level of expertise by attaining or exceeding the minimum examination grade. As members become certified at increased levels of training, a job promotion should result. This is a program of members helping members, and it has definitely aided the industry by improving the members' technical expertise.

The IEEE has been active in developing telecommunications standards and will most likely continue to do so, even though some international standards are nearly the same. The IEEE maintains many committees that work on standards and specifications. Also, several committees of the Institute work with the congressional subcommittees in advising and helping to shape legislation. As mentioned before, the IEEE maintains many chapters in foreign countries and hence has input to the development of international standards.

The SCTE is now becoming active in standards through the American National Standards Institute (ANSI). The cable television industry also has an association called the National Cable Television Association (NCTA), which represents the cable television operators and equipment manufacturers. Many people whose companies are members are also members of the SCTE. The national headquarters of this organization are appropriately located in Washington, D.C. so the cable television industry has some influence on the Washington scene. The SCTE and the NCTA has worked with the FCC in establishing system performance standards for cable television systems. Cable system leakage and proof of performance testing are some of the areas worked out among the three parties. The NCTA has published a technical operations handbook of accepted methods of testing and standard system performance specifications. Technical papers presented at NCTA conventions are available to the member organization. The SCTE has provided technical training through its several expositions and conventions, with the convention proceedings published and available to members. A series of satellite-transmitted training programs have been produced by the SCTE. The SCTE obtained satellite transmission times and produced video training programs to be broadcast on a satellite transponder. Members could either view them live or record them on a VCR for later viewing. Timely topics could be handled this way as opposed to purchasing a separate videotape from the society.

In summary, the success of the information highway depends on proper financing, availability of proper equipment, and a supply of well-prepared technicians and engineers. The federal government should provide some financial encouragement, possibly through some tax relief. The Congress should act on telecommunications reform legislation so the industry can determine its best course of action. The participating telecommunication companies in the information highway have to regard their participation as an economically viable business venture. In short, the information highway has to make money and pay a reasonable rate of return to investors.

6

The Information Highway and Global Telecommunications

6.1 Present Status of the Information Highway

The information highway is moving forward at a high rate of speed. New equipment, methods, and programs are being developed almost daily. What was a new development yesterday is improved by today's latest efforts. This is most true here in the United States. The close proximity of the United States to Mexico and Canada has allowed the technology to cross the borders. Due to similarities of the ethnic origin of the peoples, the United States and Canada have enjoyed a close relationship. The telephone systems cross-connect and operate as if there were no borders. Due to trade agreements, the United States and Mexico are also related more closely. In many cases, U.S. manufacturers maintain factories in both Mexico and Canada. As these trade relations develop, more telecommunication between these adjoining countries will be needed.

6.1.1 The status in North America

The telecommunications industry in North America is probably as highly developed as anywhere else in the world. Voice, data, and video signals cover the North American scene. Many of the satellite systems provide coverage for the United States, Canada, and Mexico. Wired systems of copper, coaxial cable, and fiber-optic cable span the area, which mostly originates in the United States. Radio and/or wireless systems are necessary to reach remote areas or for mobile applications. The telecommunications system has an enormous impact on the way business is conducted in the world today, and the economic benefits are huge.

6.1.1.1 Development of wired systems in the United States. The development of wired systems in the United States has been going on since the invention of the telephone. The telephone system consists of many small, independent telephone companies and the giant telephone companies such as MCI, Sprint, and the largest, the Bell System. This telephone system is responsible for giving the United States the best telephone technology in the world. The guiding force has been the Bell System, formed by the American Telephone and Telegraph (AT&T). Starting with the railroads, the telegraph system also grew at a rapid rate and was the first use of commercial digital transmission. Electromechanical machines took the place of the manual telegraph key and sounder, developing what was known as teletype. As demand grew, teletype was transmitted through telephone wires and was used by many banks for wired money transfers. The ASCII standard was developed from this method.

As the demand for digital data increased due to the arrival of the personal computer, it was necessary to develop some standards. The IEEE also recognized the problem and, through the Institute Study Committee, the 800 series standard was developed for LAN technology. This is often referred to as the IEEE Project 800. The efforts of the IEEE developed some ground rules for network architectures' control standards. Various layers of operations were defined. The 802 specification describes layers and sublayers of system controls. Specification 802.2 specifies the logical link control sublayer, which controls or defines the sequence of message exchanges between the communicating parties. The 802.3 is the media access control (MAC) layer and addresses the token ring or bus situation such as CSMA/CD. This sublayer also defines the Ethernet method. Specification 802.4 addresses the token-bus access-control technology, and the 802.5 covers the token-ring method. Remember that the 802 specifications have to do with control and architecture concepts, not the physical method of transmission, such as wire, cable, or fiber-optic systems. These are areas that should be addressed when data transmission is to be implemented.

The Electronic Industry Association published the RS-232 specification, which defines the transmission of pulsed electrical data representing binary-coded data words. This was important to setting some standards for the electrical interface between computers and peripheral devices. The foregoing standards were needed before chaos occurred in the industry.

The United States, operating in its free-enterprise environment, has always been quick to invent new products and techniques. New methods and devices are advertised, marketed, sold, and installed very quickly here in the United States. No wonder there are so many methods and devices in operation today. This can cause a significant problem with interconnecting systems to form the information highway. A case in point is the interconnection of networks operating in the fiber-optic realm of the fiber distributed data interface (FDDI) and compatibility to equipment operating on an asynchronous transfer mode (ATM) system. There are several systems in wide use today, and the problem of interconnection with the information highway in formidable. All of these systems require both software, usually built in, and hardware. Information or data in a variety of digital formats has to be collected, assembled, tagged, and sent via some high-speed method along the information highway. Some form of data packetizing forms the message. At the receiving end, the data often has to be unpackaged, assembled, and sent to the receiving-station destination. Routing and message length determine the charges.

No doubt there are many wired private communications systems in the United States. As mentioned previously, the telephone system has been highly developed and operates according to Bell System standards. The telephone industry uses digital methods and techniques in microwave radio links, satellite links, coaxial cable links, and fiber-optic cable systems. Voice signals are converted to digital signals that are packetized according to the DS standards and transmitted through the various physical cable systems. For fiber-optic systems, the DS-3, or higher levels, the signal is further multiplexed to optical carrier signals (OC) and can be further multiplexed to higher levels. This system is known as the synchronous optical network (SONET) and was discussed in chapter 4.

Computer data originating at a variety of commercial companies usually operates within an LAN environment. In the past, if digital data was required to be sent to a company outside or beyond the LAN, the workstation was disconnected from the LAN and the data was transferred to the receiving location through the telephone system using a modem. Specifications for such modems, originating in the United States, have been adopted, accepted, and/or modified by the international standard-setting organization CCITT. These standards are referred to as the V series standards. Table 6.1 shows V-type standards for modems given by CCITT Standards SGXVII.

The standards listed in Table 6.1 are accepted universally, making the interconnection between telephone systems and computer systems possible. Of course, the differences in commercial electrical power in foreign countries still cause problems of interfacing. All the various power adapters and converters available solve the power difference problem. The specifications in Table 6.1 solve the hardware interface problem but not the so-called language or data format issue. The CCITT X series specifications solve this problem and are given in Table 6.2.

TABLE 6.1   CCITT V Series Recommendations

V designation	Date rate	Description
0.10	n/a	Integrated circuit equipment for unbalanced, double-current line circuits
0.11	n/a	Integrated circuit modem equipment for balanced, double-current line circuits
0.13	n/a	Simulated carrier control
0.14	n/a	Start-stop mode characters transmitted over synchronous channels
0.21	300 b/s	Duplex telephone modem
0.22	1200 b/s 2400 b/s	Duplex telephone modem (using FDM method)
0.23	600/1200 b/s	Duplex switchable telephone modem
0.24	n/a	Definitions for interchange between DTE (computer) and DCE (modem) equipment
0.26	1200/2400 b/s	Standard modem for leased 4 wire circuits or 1200 b/s on standard switch telephone circuits
0.26 ter	1200/2400 b/s	Same but has echo cancelling features
0.27	2400/4800 b/s	2400/4800 for switch telephone circuits, 4800 b/s manual equalizer (bis) 2400/4800 auto equalizer (leased lines)
0.28	n/a	Electrical characteristics for unbalanced, double-current interconnect circuits
0.29	9600 b/s	Standard modem for 4 wire leased lines
0.32	9600 b/s	Standard duplex modem for 2 wire switched telephone circuits or leased 2 wire circuits
0.32 turbo	19,200 b/s 21,600 b/s	Standard switched telephone circuits (ASL)
0.32 bis	14,400 b/s	Standard switched telephone circuits
0.33	14,400 b/s	4 wire leased telephone circuits
0.34	28,800 b/s	Standard switched telephone circuits
0.35	48,000 b/s	Using 60 kHz–108 kHz group-band circuits
0.36	64,000 b/s	Synchronous transmission using 60 kHz–108 kHz group band circuits
0.37	Greater than 72,000 b/s	For synchronous transmission on 60 kHz–108 kHz group band circuits
0.42	57,600 b/s	Error control/data compression. Asynchronous to synchronous conversion
0.52	n/a	Loop testing device specifications for modems
0.110	n/a	DTE support using V series interfaces by an ISDN (Integrated Services Digital Network)
0.120	n/a	Support for ISDN DTE with V series interfaces and statistical multiplexing
0.230	n/a	General interfaces for data communications

TABLE 6.2 CCITT X Series Recommendations for Data Communications Interfaces with Public Networks

X designation	Device interface	Description
0.20	*DTE/DCE	Start-stop transmission telephone type networks
0.20 bis	DTE/DCE	Start-stop transmission, asynchronous duplex V series modems
0.21	DTE/DCE	Synchronous operation, DTE designed for synchronous V series modems
0.22	DTE/DCE	Multiplex interface for classes 3 to 6
0.24	DTE/DCE	Definitions for interchange circuits on public-data-type networks
0.25	DTE/DCE	Packet mode or dedicated lines of public data network
0.26	N/A	Electrical specifications for integrated circuit equipment using double-current, unbalanced lines
0.27	N/A	Electrical specifications for integrated circuit equipment using double-current, balanced lines
0.28	DTE/DCE	Start-stop made DTE access a Packet Assembly Disassembly (PAD) facility in a public data network
0.29	N/A	Control information procedures for a PAD facility in a public data network
0.30	N/A	X.21 & X.21 bis support-based DTEs on an Integrated Services Digital Network ISDN
0.31	N/A	Support for packet mode equipment on an ISDN
0.32	DTE/DCE	Interface for DTE operating in packet mode for connection to a packet switched network, a public switch telephone network or a circuit switch public data network
0.75	DTE/DCE	Interfacing between packet switched networks

*DTE–Data Terminal Equipment (work station, P.C. etc) DCE–Data Communication Equipment (Modem or Interface between Communication network and data terminal).

The problem of the physical connection was formed first in the United States by the Electronic Industry Association (EIA). This resulted in the industry acceptance of the RS 232C standard. Many of these standards have been accepted internationally. Some EIA interface standards are given in Table 6.3.

The many privately owned wired or cable systems belonging to corporations, schools, colleges, and municipalities range from old to new, and rudimentary to ultramodern. Connections of these systems to the information highway will be varied. This is an area where new interface products will be required. No doubt there will be new companies that will be formed to become involved with interfacing.

TABLE 6.3 EIA Specifications for Data Interface Equipment

Name EIA standard	Description
RS-232C	DTE/DCE interface for serial binary data at rates greater than 20 Kbps
RS-366A	Interface specifications between DTE and automatic telephone calling equipment
RS-422	Electrical characteristics for balanced voltage digital circuits
RS-423	Electrical characteristics for unbalanced voltage digital circuits
RS-449	Specifications for 37-position and 9-position DTE/DCE equipment using serial binary data interchange at greater than 2 Mbps
RS-496	Interface between DC and PSTN

Rates and interconnect line charges will be as complicated as they are today. A new class of communication consultants with a huge database of routes and charges will be needed by many small business customers to determine the lowest-cost methods that result in the best service. This is being done today for telephone customers when the type and time of the telephone services for a client are studied. Recommendations as to the telephone carriers and equipment are made to a client for improved service and lower costs. Many businesses are finding the results well worth the consulting fee.

6.1.1.2 Development of wireless systems in the United States. Wireless communications systems are growing in the United States at a rapid rate. Probably the most visible is the personal pager and the portable cellular telephone. The electronic manufacturing industry in the United States is very resourceful. New products and devices are introduced rapidly, as evidenced by the advertising media. Electronic devices are becoming smaller and less expensive. Packet pagers activated by low-power radio cells alert the carrier of a telephone call, and today many even indicate the telephone caller's number on a liquid crystal display (LCD). This service is lower in cost than the portable cellular telephone. The cellular telephone employs a wireless pickup into the telephone network. Some cable television operators are adapting their cable system to be able to pick up cell signals and connect into the telephone network at their head-end facilities, thus becoming a cellular telephone operator.

It seems it was only a few years ago that wireless or radio-type systems in the United States consisted of radio and television broadcasting, commercial communications systems for air and marine traffic, amateur radio operations, and CB radio. Now satellite broadcast television systems, satellite global positioning systems, and satellite commercial business communications are being used on a daily basis. The newest version of satellite television is the digitally transmitted direct satellite broadcast (DBS) system, which uses a very small amount of exterior equipment.

Since much of the commercial development of wireless systems was performed here in the United States, that's where many standards were forthcoming. The National Association of Broadcasters was an important influence in setting broadcasting standards. The Federal Communications Commission (FCC) has, through its rules and regulations, set the standards for the broadcasting industry in the United States. Essentially, the FCC has jurisdiction over all the public wired telephone networks and all radio transmissions.

From a purely theoretical point of view, the radio transmission spectrum bandwidth covers from the very low frequency 300 kHz to 40 GHz, according to the international trade union (ITU). One could say that this relates to a total transmission bandwidth of approximately 40 GHz. Fiber-optic cable can operate in the region between frequencies corresponding to a wavelength of 1550 nm to the shorter wavelength of 850 nm. This relates to frequencies of 1.94×10^{14} Hz @ 1550 nm and 3.53×10^{14} @ 850 nm, resulting in a bandwidth of approximately 1.6×10^{14} Hz or 160,000 GHz. Therefore, comparing the bandwidth for a fiber-optic system to a radio system, there is a ratio of 4000 to 1. Also, from a theoretical standpoint all of today's communication could be transmitted through fiber-optic cable systems with room to spare, thus leaving to radio what it does best. Protection of the radio spectrum is very important to the information highway. Radio techniques will be utilized where cable systems do not exist, such as for space travel, air and sea transportation uses, and highly mobile communications such as personal communications.

6.1.2 Present status in the world marketplace

Telecommunication systems in the rest of the world range from excellent to poor. Systems in Japan and much of Europe are as good as what we have in the United States. However, there are horror stories. In France the telephone directory will be available on video phone. However, to get a telephone installed takes a very long time. In Sweden the Ericson Company manufactures highly developed telephone and telecommunications equipment, and it is marketed here and in Europe. Fiber-optic cable systems span much of Europe, including submarine cables connecting island countries and continents. Much of this equipment is manufactured and often installed by companies in the United States.

6.1.2.1 Wired systems in foreign countries. The telephone systems in Europe follow the CCITT standards, which had their source from standards developed by the Bell Telephone Company here in the United States. Maintaining compatibility meant that there would be a certain amount of equipment interchangeability. Thus equipment manufactured in the United States could be used in Europe, and vice-versa.

The trade agreements between many countries of the world have resulted in the expanded telecommunications systems necessary to conduct international business. As discussed in a previous section, fiber-optic long-distance submarine cable systems connect many island countries together. Multimode fiber-optic cable with a larger cone of acceptance allows the injection of higher-power laser sources to drive the cable. Due to the effects of dispersion, lower data rates are necessary to achieve longer distances. Optical amplifiers along the submarine cable route are necessary to obtain the long-distance spans. Metallic cable members are employed to power the cascades of optical amplifiers. The digital data rates are still very large compared to metallic submarine cables.

Third-world countries are severely lacking in the telecommunications area. The more economically developed countries have a more highly developed telecommunications system. Many of the so-called third-world countries have little or no telephone system. In heavily populated areas, there might be a telephone system, but many are not connected. Satellite-based systems are often the best choice for improving the telephone interconnect situation.

Many countries in Europe and the Orient also have cable television systems. Here was a clear-cut case of an industry taking off with little or no standards. In England a hybrid television system called the rediffusion system was used at one time. Generally, the development of cable television systems in Europe took place at a slower rate. This is also true of the television broadcasting industry. However, as time went by, the television and cable television industries developed. The television transmission standards are quite different from the NTSC system here in the United States. There are two standards in Europe called PAL and SECAM. The PAL system (phase alternating lines) is most like the NTSC system used in the United States, Canada, Mexico, and Japan. The PAL system employs a single-phase and amplitude-modulated subcarrier superimposed on the luminance signal. The PAL method is used in Europe, with the exception of France. The SECAM method (sequentiel couleur avec memoire) used in France uses two frequency modulated subcarriers. Both the PAL and SECAM methods for color television transmission use slightly more bandwidth than the NTSC method of 4.2 MHz. The PAL method used 5–5.5 MHz, and SECAM requires 6.0 MHz. UHF transmission is used by most European television stations.

The European countries that were affected by World War II often have a more improved telecommunications industry, simply because it had to be reconstructed after being destroyed during the war. Gone were old telephone cables and antiquated equipment. These were replaced by second-generation equipment.

6.1.2.2 Wireless systems in foreign countries.

Europe indeed does have a satellite system in place that provides service throughout much of Europe and Africa. An active satellite telecommunications system called Asia Sat is also developing in Asia and was started in 1990 with its first satellite, Asia Sat 1. This satellite actually had humble beginnings. It was earlier known as Western Unions Westar 6 and was stationed in a useless low orbit. It subsequently was rescued by a space shuttle crew, and after being rebuilt it was sold to Asia Satellite Telecommunications Company Ltd. They renamed it Asia Sat 1 and relaunched it in April of 1990. It has performed well, and all transponders are full and providing telecommunications and television programming to many Asian countries. Asia Sat 2 is scheduled to be launched during mid-1995 and is a high-power satellite. Asia Sat 1 was reworked by Hughes, and Asia Sat 2 is being manufactured by Martin Marietta Corporation. Asia Sat 1 has a C-band system of transponders, and Asia Sat 2 has 24 C-band transponders and 9 Ku-band transponders. The frequency ranges for the satellite bands are given in Table 6.4.

Satellite technology at present is an extremely important carrier of telecommunications. Properly positioned and orbiting satellites can provide communication coverage for any country in the world. The International Telecommunication Union (ITU), which is a member organization within the United Nations (UN) is involved with the common telecommunications agreements among the member nations. The former CCITT and CCIR organizations are now part of the ITU.

The ITU has divided the world into three regions for allocating frequencies for various telecommunications services, including satellite services. The ITU Region 1 consists of Europe, the Scandinavian countries, Africa, the Middle East, Russia, and Siberia. Region 2 covers North, South, and Central America, Canada, and Greenland. Region 3 essentially covers the rest of the world, consisting of India, Australia, New Zealand, Phillipines, Japan, China, and other Asian countries.

TABLE 6.4 Satellite Transponder Bands

Band letter	Frequency band (GHz)	Remarks
L	1–2	L & S bands also known as higher UHF band
S	2–4	Same as above
C	4–8	C, L and S, C band, commonly used
X	8–12	Usually used by government/military
Ku	12–18	Rain can cause problems. Ku band needs more power
K	18–27	Rain causes large signal attenuation
Ka	27–40	Rain causes large signal attenuation
V	40–75	40.5–42.5, 42.5–43.5, 47.2–50.2 usually used
W	75–110	Not commonly used
MM	110–300	Not commonly used

International Conferences sponsored by the ITU are held every two years and are referred to as World Radio Communication Conferences. Delegates from the countries residing in these regions discuss, study, and make recommendations for the purposes of frequency allocations for telecommunications services and equipment. The WRC conference is prepared by many groups and committees prior to the actual convention. Many of these committees solicit opinions and suggestions from the regional telecommunications system operators and equipment manufacturers. Reuse and sharing of the frequency bands is an important topic to be settled and agreed upon for proper interference-free radio communication services operating within the three regions. Many microwave radio systems operate within the same radio bands, as do satellite services. Therefore the operating frequency bands have to be separated so each system can operate without interference.

6.1.3 Standards and standard-setting organizations

The telecommunications industry as operated in many countries in the world today ranges from being very highly regulated to not regulated at all. This depends on whether the country has a well-developed economy or not. Most countries have a government commission, department, or division to regulate and oversee the telecommunications activities. In the United Kingdom, the post office was the designated telecommunications department. Evidently, since the delivery of printed messages between parties (mail service) is regarded as a form of communication, this department was a natural to administrate all communication activities for the country. With the development of wireless systems, i.e., radio communications as we know it, regulation in Europe became necessary because radio waves know no borders. Frequency allocation was and still is important for high-power radio communications systems operating throughout the world. It is becoming more and more important to develop world standards for the telecommunications industry. Setting such standards will allow improved sales and service for telecommunications equipment sold throughout the world. Since the formation of the United Nations following World War II, the International Telecommunications Union (ITU) has developed as the administrator of the world telecommunications agreements.

6.1.3.1 Standards in the United States. Telecommunications standards are essentially the laws, rules, and regulations observed by the telecommunications industry. The regulating agency for all telecommunications systems operating in the United States is the Federal Communications Commission. The FCC was a result of the Communications Act of 1934. The FCC has formed rules and regulations that have grown in number since the commission has been in existence. The two types of communications systems are wired and wireless, and the FCC has jurisdiction over both of them.

The available frequencies used for wireless services have grown immensely over the years. What was once regarded as ultrahigh-frequency (UHF) is certainly not very high by today's standards. Wireless emissions are specified as to frequency use, and bands of continuous frequencies are specified as to the use and type of wireless services. For example, certain frequency bands are allocated for radio and television broadcasting, business band radio services, aircraft and marine radio and navigational services, radar, satellite telecommunications, etc. The frequency boundaries for the allocated bands and the actual carrier frequency accuracy for each assigned carrier frequency are written into the volumes of rules and regulations covering these services. Signal quality, radio station signal, service-area patterns, and testing and measurement specifications appear in the rules and regulations.

Wired systems have exacting signal-quality standards and test and measurement specifications. Such signal problems as echo, loudness, and cross talk for telephone systems are covered in sections of the wired telephone section of the rules and regulations. The FCC also has jurisdiction over rates and charges for both types of systems.

The sole purpose of the commission is "for the public's best interest or public good." The guarantee of high-quality, minimum services at appropriate rates and charges for the telecommunications system is the goal of the commission. Often the FCC has to act as an intermediary in disputes between rural companies and/or the public. Now that digital systems and fiber-optic cable systems have been developed, the commission's work load has increased significantly. Today the industry is patiently waiting for changes in the rules to allow growth of the information highway. These changes will be forthcoming when Congress passes legislation pertaining to the rules and regulations of the FCC.

One of the newest and latest of the wireless systems is the personal communication system (PCS), which includes cellular and microcellular telephone services and paging services. Cellular service areas are divided geographically. Two different classes of areas result. They are the metropolitan service area (MSA) and the rural service area (RSA). The FCC has allocated the top (most heavily populated) areas to A-type nonwired system operators and to B-type local telephone exchange operators. Therefore two companies can operate in a market area, thus encouraging competition, which could result in serving the public's interest.

These top market areas were auctioned off to the highest bidder. The remaining MSA areas (31–120) were divided between nonwired system operators through mergers and alliances before putting them into a lottery. The RSA areas with the remaining MSA areas were or are allocated by lottery. Much of the public is anxiously awaiting the development of this section of the telecommunications industry. It is these local wireless systems that are going to pick up large amounts of voice or telephone traffic and transfer them to the information highway. Laptop and palm-held portable computers using wireless modems can provide data communications from the local cells to anywhere in the world travelling the information highway.

Other organizations operating in the United States that have and are providing an important influence on the telecommunications industry are the IEEE, the EIA, and ANSI. All have addressed certain areas of the industry. The IEEE has contributed to the digital data area with the 800 series of recommendations known as Project 800. Network configurations and software protocols are contained in the 800 series of IEEE recommendations. The EIA has made recommendations as to the electronic equipment referred to as hardware. The RS specifications as to connector types, pin assignments, and wire functions are found in the EIA recommended specifications. In the United States, the EIA is concerned with electronic component and equipment standards and is an important influence in the electronic industry. The EIA was formerly known as the Radio-Electronics and Television Manufacturers Association and has its headquarters in Washington, D.C. Many areas addressed by the EIA are also covered by ANSI and are for all practical purposes the same. ANSI was formerly known as the U.S.A. Standards Institute and has its headquarters in New York City. ANSI and EIA cooperate in many areas of the electronics industry. The U.S. military has played an important part in standards and specifications, principally as a part of the equipment procurement process. The MIL-STD (military standard) for components, equipment, and the testing procedures used to prove standards is found in the volumes of MIL standards. For electrical equipment and machinery, the National Electrical Manufacturers Association NEMA is very active and also maintains its headquarters in New York City.

The federal government formed the National Bureau of Standards (NBS), now known as the National Institute of Standards (NIST), and is concerned with physical standards for such quantities as time, weights, and measures. NIST maintains a series of radio broadcast stations that transmit the standard time, both in code and by voice. The carrier frequency is maintained to the highest standards of accuracy so it can function as a standard frequency. Frequency accuracy is one part in 10^{12} and is precisely monitored by NIST. Electrical standards maintained by NIST are the standard unit of resistance (the ohm), the standard for capacitance (the farad), and various other electrical quantities such as the volt and ampere, to name a few. NIST has its headquarters in Boulder, Colorado. Long-term signal variations due to atmospheric conditions and time of day are studied at NIST, and predicted values of transmission loss at various frequencies are made. Studies such as this have been helpful in the progress of radio communications throughout the world. The existence of the NBS and now NIST has been a considerable benefit to the scientific world.

6.1.3.2 Standards in Europe. The European continent has many adjoining countries who are making necessary telecommunications agreements. Compatibility of telephone and telegraph systems was necessary for communication between countries. There are two separate groups presently working to make recommendations on wired system standards and wireless system standards. The CCITT group (International Telegraph Telephone Consultative Committee) administers the area of wired systems and CCIR (International Radio Consultative Committee) administers the wireless area. Both of these organizations operate under the auspices of the ITU. The recommendations of these groups are updated, usually following a world meeting or convention, often referred to as a plenary assembly. These occur every three or four years. The CCITT publishes volumes of recommendations referred to as a color-coded book. The *Red Book* was published in 1960, the *Blue Book* in 1964, the *White Book* in 1968, the *Green Book* in 1972, the *Orange Book* in 1976, and the *Yellow Book* in 1980, to name a few. The CCIR uses no color coding of the volumes representing the agreed recommendations. The main concern is the compatibility of message traffic being maintained between both wired and wireless systems. Most likely the information contained in these volumes will be available in a library of CD-ROMs, if it isn't already. Since the United States and most countries in the free world are UN members and have delegates in the ITU, the CCITT and CCIR organizations' worldwide telecommunications will and should be attainable.

6.2 Integrating the Global Information Highway

Integrating and administrating the operation of the information highway on a global basis is an extremely difficult and involved task. So many forces are at play—technical, economic, and governmental. These forces can be both cohesive and divisive. It is incredible how far this effort has come in the world today in light of the world situation, with its areas of conflict and unrest. However, it is again a two-way street. The advances in telecommunications that have caused the increased speed of messages can be a benefit and a problem as well. For example, if a border conflict between countries erupts and the world hears about it nearly instantaneously, certain countries can act to take advantage of the situation, while some might try to act out of compassion. Still this isn't a communication problem, even if communications indeed has an effect. In the long run, most countries benefit economically and cooperatively as the telecommunications system improves.

6.2.1 System forces

The telecommunications industry is often regarded as a force acting on society. Instantaneous knowledge about world affairs and situations is indeed important in how government and industry reacts to them. The old saying is "bad news travels fast." Today, bad news travels almost instantaneously. In the business world, associates can act almost as one with a highly developed telecommunications industry. Aid between foreign countries can be enhanced by the exchange of medical, business, and financial information. In other technical areas, people are finding and using the telecommunication system's high information capacity to enhance and increase their business and/or improve the way they do business. Some workers are finding that they can do some of their work at home using a personal computer setup. Thus the trip of going to and from work is saved, and the time spent at home is spent on productive work. This might also allow a single parent to work at home and care for children. Such a situation has an effect on the transportation system and/or parking facilities. A short examination and discussion of these effects and forces will demonstrate the importance of the development of the information highway.

6.2.1.1 Economic forces/benefits. For some people, the greatest benefit the information highway has to offer is economic. As said many times by many people, there is money to be made here. A highly developed telecommunications system will have an enormous effect on the way business is carried out in the world and will have a large economic impact. There will be a plus and a minus side to the economic scene. Certain types of jobs will disappear and others will appear with a cause-and-effect scenario. The case where some people will work part of the time at home will mean that transportation, parking, commuter shopping, etc., will be affected.

More efficient ways to conduct business will be forthcoming, with resulting cost savings. Brochures and advertising pamphlets introducing new products can be sent by sales personnel to prospective customers using the information highway. The postal system will most likely experience fewer advertising pieces of mail. The information highway is being used today by many marketing and sales organizations. Ordering, purchasing, and paying for merchandise is presently being carried on by the various networking systems that are already in place. As the information highway becomes more developed, more telecommunications sales and marketing will be done.

One of the biggest benefits of the information highway will be in the new field of telemedicine. The transfer of medical information in the form of patient's medical records, X-rays, and diagnostic imaging scans will be carried out simply and effectively from one part of the world to another almost instantaneously. Medical consultation can and will be carried out on the information highway. The quality and cost of health care could be greatly improved. The day might come where certain surgical procedures could be performed using robotic systems controlled by a surgeon many miles away. This is probably scary to consider. However, it is indeed possible.

Manufacturing can also benefit from a highly developed telecommunications system. Many manufacturing plants operating today use robotics and automated methods wherever possible. Scheduling factors such as plant supplies, equipment, and delivery charges can affect plant operation and can be controlled remotely by a high-speed telecommunications network. Costs and controls in the manufacturing process are extremely important for highly efficient and profitable plant operations.

The transportation industry has always been dependent on telecommunications to some degree. Aircraft and marine operations have used radio voice communications for scheduling and safety reasons. Also, radio and radar operations and navigational and positioning electronic systems provide wireless telecommunications for the transportation industry.

Any country's economy and well-being can be improved by a highly developed telecommunications system. Trade, transportation, and diplomatic issues can be taken care of quickly by taking advantage of a good communications system.

6.2.1.2 World peace and security elements. World peace and security definitely can be affected by high-speed, reliable communications. Prompt reaction to emergencies, threats, and/or military actions can be resolved with the aid of decent telecommunications. At best, problem situations could be foreseen and resolved before the problem worsens or increases to dangerous proportions. In the past, military operations were large users of telecommunications facilities. Telephone systems were the best and first systems to be developed and were relied upon heavily during periods when the country was at war. Voice and teletype services were used during both World War I and World War II. Now the telecommunications that is in place today and improving constantly provides voice, video, and data communications throughout the world. Surveillance and monitoring of sensitive government-controlled areas use various types of telecommunications services. An encryption method is used to control access of information to authorized offices. Several types of encryption methods have been developed to provide secure communications between authorized parties. The cable television industry uses such methods to prohibit unauthorized reception of satellite services and unauthorized cable services to subscribers. The direct satellite broadcast service (DBS) uses similar techniques.

The personal communications services (PCS) have some systems that provide some method of signal security. Many of these methods had their beginnings with some of the governmental telecommunications efforts.

6.2.2 Control of the international information highway

The control, regulation, and maintenance of the information highway will be a formidable task at best. Since many governments will be involved, it is probably wise to consider how to handle it properly here in the United States before taking the global scene into account. The main areas of concern are the areas of the system itself, the rates, tolls and charges, and the selling, marketing, and promotion. Since many separate companies are involved with the many sections of the system, each area of concern will be addressed separately.

6.2.2.1 System and technical operations. The system is and will consist of many separate contributions to the overall network, which will transmit and receive data for the end customers or subscribers. Metallic wire, coaxial cable, and fiber-optic cable supported by wireless systems all contribute to the information highway telecommunications network. Naturally, each company will maintain its own section of contributing plant. But some means of testing and controlling the end-to-end signal quality is going to have to be resolved. If an end-to-end problem occurs, then the problem section has to be identified and corrected. For a highly reliable system, such testing has to be continual so when a problem does occur, the message traffic can be rerouted through redundant paths. This rerouting has to be self-healing and completely automatic.

Certain types of network topology are better than others for completely self-healing communications. For example, the star-type topology has more redundant routes. So if an area of the network suffers an outage, the network diagnostic system senses the problem and reroutes traffic around the affected area. This is very similar to an actual highway accident, bridge outage, or flooding disaster where traffic is immediately rerouted. It should be apparent that the network should be a smart network carrying diagnostic messages along with customers' message traffic. Also, the network should have enough spare capacity to ensure that it will not become overloaded during outages. This spare capacity should be designed into the system, taking into account the statistical nature of outages. The spare communications capacity built into a network should not be used for increased customer use at the expense of reliability. A properly designed communications network should have enough capacity to allow an increase of business with a proper margin for automatic outage rerouting.

Since so-called smart networks use computer-controlled diagnostics and switching methods, protection of the software against willful tampering or espionage is extremely important. Protection against software viruses, computerized thieves, and independent hackers is necessary to keep the communications network safe and operating properly.

A lot of diagnostics and antivirus software is appearing on the market, along with microprocessor-controlled communications diagnostic and monitoring instrumentation. The Hewlett Packard Company has devoted a whole separate division to communications diagnostic instrumentation. Instruments have been developed to work in a variety of network environments such as ATM and SONET systems. Many other instrument manufacturers make instruments to test and monitor communications facilities, but these companies do not devote an entire division to the endeavor.

From a personal standpoint, there is a developing need for more highly trained technical people in the telecommunications area. The RBOCs are probably a bit better off than some of the cable television companies, due to a strong in-house company-operated training program. Colleges, universities, and technical vocational schools had better see the light and develop some programs before the need gets too great. There will be a need for telecommunications personnel at all levels: engineers, technicians, network managers, and administrators.

Trained personnel is the main ingredient for any successful business. A case in point is the cable television business. Since there was no training ground for this business, people came from all areas of the business and technical world. Some technicians came from the radio and television repair businesses, the telephone industry, electrical contractors, or the power companies. Managers were formerly insurance or banking people or local merchants. The more fortunate had some accounting and/or sales experience. There were also very few available books on cable television technology back in the very beginning. The equipment manufacturers were quick to publish technical papers along with their advertising brochures and catalogs. Also, some of the larger manufacturers conducted area training seminars, which provided much-needed information to the new industry. The technicians and engineers, mostly out of necessity, banded together and founded several organizations, which were the National Cable Television Association (NCTA), the Society of Cable Television Engineers (SCTE) now known as the Society of Telecommunications Engineers, and the Cable Television Association (CATA). Several other societies pertaining to sales, marketing, and management were also founded. It was the efforts of these societies and the equipment manufacturers' training programs that caused this industry to become fully developed.

On the international scene, the information highway will be controlled and administered mostly through the United Nations' International Telecommunications Union (ITU). The responsibilities of the CCITT and the CCIR are rolled in under the ITU. To speed up the study work on standards and specifications, the ITU is expanding its computer facilities. The subcommittee's work will be entered into a database, so after the plenary session the published results will be available much sooner. It used to take four years to produce the records of each plenary session. With the new system, results should be available in 18 months. The ITU has dropped the CCITT and CCIR acronyms and instead has developed ITU sectors responsible for separate areas of telecommunications topics. It is expected that this streamlining of the ITU will allow it to keep up with the explosion of the world telecommunications industry.

6.2.2.2 Rates, charges, and tolls. The control of costs, rates, tolls, and charges is extremely important to most everyone involved. Such charges to customers or subscribers, as the case might be, is under control of the FCC, and the legislation was passed by Congress here in the United States. The rate structure is set by common agreements between bordering countries, i.e., Mexico, Canada, and the United States. On the international scene, the ITU reviews the rate structures between the member countries. As far as most communications systems are concerned, the originator of the connection (the calling party) pays the cost of the communication. The telephone exchange carrier or other carrier supplying service to the originator of the call or service collects the total charges based on the rate and time length of the connection and distributes the funds for the various suppliers along the connecting route. This method has been used by the various telephone companies throughout the world and so far seems to work well. Although it sounds simple, the differences in the exchange rates of the various currencies and the various costs and charges complicate the procedure. The increased use of computers and computational equipment has eased and simplified what was once a labor-intensive system.

As the networks get smarter, more efficient routing will result, which can lower costs. Such added features as automatic connection timing and memories for called- and calling-party numbers are all a benefit of smart network switching. Remember that so-called smart networks operate under some programmed computer or processor system, and bugs or software problems can be equally as important as hardware malfunctions. Many exchange carriers have equipment on hot standby, and diagnostic software continually checks and tests. Using fiber-optics, the SONET system has built circuit redundancy and software redundancy into the network. Continual in-service testing is carried out, and if problems occur, alarm signals are activated and message traffic is rerouted. SONET test instruments are then used to locate the faulty equipment, which will be either repaired or replaced. As mentioned before, the SONET system will become the main toll road on the information highway.

6.2.2.3 Marketing, promotion, and sales. As in most business enterprises, selling the service or end product is very important to the venture's economic outcome. The information highway can be very confusing and new to many prospective users. Many types of new businesses face the prospect of making sales and marketing decisions that can affect the success of the business enterprise. Fortunately, since the telephone industry has been actively providing communication services to the business and private communities, they already have sales and marketing departments in place. Since the benefits of the information highway can provide important services to many businesses, these benefits have to be marketed to prospective customers.

Many of the large, successful companies operating in the United States already know of and are using available communication facilities. As the information highway improves with proper deregulation and an improved economic climate, the benefits have to be properly advertised and marketed to present and prospective customers. The information highway will be fully realized once such benefits as information transfer rates and cost factors improve and the facts are made known through the communications media, word of mouth, and normal advertising methods. The business world will then be able to operate with incredible efficiency. Hopefully, the resulting cost of service reductions will be passed on to consumers.

A

Wire Loop Resistance

From physics the resistivity P for soft drawn copper wire is given as 1.724×10^{-8} Ω.m.
Resistance of copper wire of length L and cross-sectional area A is given by

$$R_w = \frac{PL}{A}, \text{ which can be written as}$$

$$R_w = \frac{PL}{\pi\left(\dfrac{d}{2}\right)^2} \text{ where d is the diameter of the wire in inches}$$

$$P = 1.72 \times 10^8 \Omega.\text{m} \times 39.37 \text{ in/m} = 67.9\Omega \text{ in} \times 10^{-8}$$

$$R_w = \frac{67.9\Omega.\text{m} \times L \times 10^{-8}}{3.14 \quad \dfrac{d^2}{4}} = \frac{4 \times 67.9}{3.14} \quad \frac{L \times 10^{-8}}{d^2}$$

$$R_w = \frac{86.5 L \times 10^{-8}}{d^2} \qquad R_w = \frac{86.5 \times 10^{-8} \ \Omega.\text{in L}}{d^2 \ \text{in} \times \text{in} \times 1\text{ft}/12 \text{ in}}$$

$$R_w = \frac{86.5 \times 12 \times 10^{-8}\Omega \, L}{d^2 \text{ft}} = \frac{10^{38} \times 10^{-8}\Omega \text{ft} \, L}{d^2 \text{ft}}$$

$$R_{LOOP} = 2R_w \text{ for a mile of 5280 ft}$$

$$R_{LOOP} = \frac{1038 \times 10^{-8} \ \Omega}{d2 \ \text{ft}} \times 5280 \text{ ft/mi} \times 2$$

$$= \frac{1.038 \times 10^{-5} \times 5.28 \times 10^3 \times 2}{d^2} = \frac{10.96 \times 10^{-2}}{d^2}$$

$$\boxed{R_{LP}/\text{mi} = \frac{0.1096}{d^2}} \quad \text{where d is the diameter in inches.}$$

For AWG-#22 wire d = 25.35 mils where 1 mil = $\dfrac{1}{1000}$ in = 1×10^{-3} in

$$R_{LP} = \frac{0.1096}{\left(25.35 \times 10^{-3}\right)^2} = \frac{0.1096}{643 \times 10^{-6}} = 0.17 \times 10^3 = 170 \ \Omega/\text{m}$$

$$R_{LP}/1000 \ \text{ft} = \frac{170 \ \Omega/\text{mi}}{5.28 \ \text{mi}/1000 \ \text{ft}} = 32.2 \ \Omega/1000 \ \text{ft}$$

B

Velocity of Propagation

Velocity of propagation

$$\text{Velocity} = \frac{\text{distance}}{\text{time}}$$

Velocity of propagation of electrical energy along a transmission line of length D meters is given by

$$v_P = \frac{D}{\sqrt{LC}}$$

L = inductance of line henrys
C = capacitance of line farads
v_P = meters per second

Characteristic impedance of a transmission line
A schematic diagram of a section transmission line is shown below

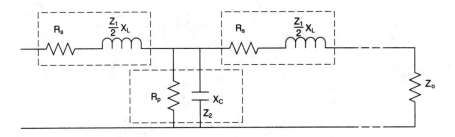

R_S is ac series resistance in ohms caused by wires.
X_L is series inductive reactance caused by twisted wires.
R_P is insulation leakage resistance between conductors.
X_C is capacitive reactance caused by capacitance between conductors.

It can be shown that $Z_0 = \sqrt{Z_1 Z_2}$

Where Z, is the impedance of 2 (R_S in series with X_L)

Also, it can be shown that if $R_S = 0 - R_p = 00$ for the lossless transmission line

$$Z_0 = \sqrt{\frac{L}{C}}$$

Therefore as series inductance is added by loading the pair, the characteristic impedance of the line is increased.

Transmission Line Calculations

Given a transmission line shown in the diagram below

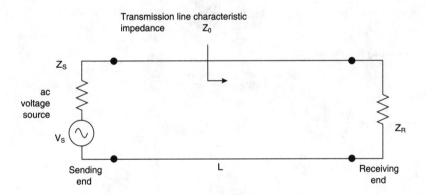

If Z_S and Z_R do not equal Z_0, voltage and current reflections will result along the transmission line length L.

The reflection coefficient K is defined as the magnitude of the reflected voltage wave to the incident voltage wave.

$$K = \frac{|V^-|}{|V^+|}$$

V^- RMS voltage of the reflected wave, reflected from the mismatch impedance.

V^+ RMS volts of the incident wave going toward the mismatch impedance.

Example: If $V^- = 0$ V there is no reflected wave and $K = 0$.

K in terms of the impedance mismatch is represented as:

At the sending end

$$K_S = \frac{Z_S - Z_0}{Z_S + Z_0}$$

At the receiving end

$$K_R = \frac{Z_R - Z_0}{Z_R + Z_0}$$

Example: For a perfect impedance match $Z_0 = Z_R = Z_S$

SO $K_S = 0$ K_R 0   No reflections

Since V^+ is the incident voltage wave traveling down the line & V^- is the reflected wave traveling up the line caused by the mismatch of the receiving end impedance. V MAX occurs when they reinforce each other to cause a voltage maximum.

|ABSOLUTE|
VALUE   a) $|V_{MAX}| = |V^+| + |V^-|$ Cancel

And   b) $|V_{MIN}| = |V^+| + |V^-|$ V^+ & V^- each other.

The voltage standing-wave ratio is defined as:

$$VSWR = \frac{|V_{MAX}|}{|V_{MIN}|}$$

Substituting a.) & b.) in above

$$c. \quad VSWR = \frac{|V^+| + |V^-|}{|V^+| - |V^-|}$$

Multiplying c.) by $\dfrac{|V^+|}{\dfrac{1}{|V^-|}}$

$$VSWR = \frac{1 + \dfrac{|V^-|}{|V^+|}}{1 - \dfrac{|V^-|}{|V^+|}} \qquad \text{and} \quad K = \frac{|V^-|}{|V^+|}$$

$$\boxed{VSWR + \frac{1 + K}{1 + K}} \qquad \text{K can be } K_R \text{ or } K_S$$

Return loss is a measure of the incident wave power to the reflected wave power mathematically

$$\text{Return loss} \quad RL = 10 \log \frac{|P^+|}{|P^-|}$$

$$\text{and } |P^+| = \frac{|V^+|^2}{Z_o}, \quad |P^-| = \frac{|V^-|^2}{Z_o}$$

Substituting

$$\text{RL } 10 \log \ \frac{\dfrac{|V^+|^2}{Z_o}}{\dfrac{|V^-|^2}{Z_o}} \ = 10 \log \ \frac{|V^+|^2}{|V^-|^2}$$

$$\text{RL} = 20 \log \ \frac{|V^+|}{|V^-|} \quad \text{Since } K = \frac{|V^-|}{|V^+|}$$

$$\boxed{\text{RL} = 20 \log \ \frac{1}{K}}$$

Return loss is the difference expressed in dB between the incident and reflected waves along a transmission line.

Coaxial Cable System Powering

A section of cable plant is shown below with resistances of cable sections shown and the distances between amplifiers noted.

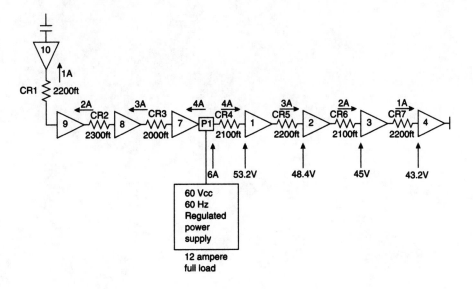

—||— Power block signal pass.

Cable section	Length 1000 ft	Resistance ohms*	Current in amps	Voltage drop † volts †
CR1	2.2	1.8	1A	1.8
CR2	2.3	1.8	2A	3.6
CR3	2.0	1.6	3A	4.8
CR4	2.1	1.7	4A	6.8
CR5	2.2	1.8	3A	4.8
CR6	2.1	1.7	2A	3.4
CR7	2.2	1.8	1A	1.8

Cable loop resistance is 0.8 ohms per thousand feet

The amplifiers each draw 1 ampere each and minimum allowable amplifier voltage is 40 volts

Power supply delivers 8 amps

* $0.8\,\Omega/1000^{1} \times$ length in thousands of feet.
‡ column 4 × column 3

Now the voltage at each amplifier location can be calculated and labeled on diagram.

Amp 1 60V–6.8 = 53.2V   amp 2 53.2–4.8 = 48.4V   amp 3 48.4–3.4 = 45
Amp 4 45V–1.8 = 43.2V   amp 7 no voltage drop.
Amp 8 60V–4.8 = 55.2V   amp 9 55.2–3.6V = 51.6V   amp 10 51.6–1.8 = 49

All amplifiers have proper voltage, and power supply is only 75% fully loaded.

Broadband Noise Combining

Noise-combining effect in single cable
reverse cable television systems

For a subsplit reverse system where carrier C_1 is at channel T-7 (7.0 MHz) and C_2 is at channel T-8 (13 MHz). And these channels are combined in a splitter and a directional coupler with equal noise levels.

Channel T-7 $C_1 = +20$ dBmV carrier level $N_1 = -25$ dBmV noise level

Channel T-8 $C_2 = +20$ dBmV carrier level $N_1 = -25$ dBmV noise level

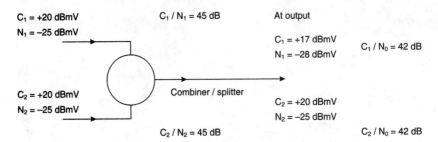

$C_1 = +20$ dBmV
$N_1 = -25$ dBmV

$C_1 / N_1 = 45$ dB

At output

$C_1 = +17$ dBmV
$N_1 = -28$ dBmV

$C_1 / N_0 = 42$ dB

$C_2 = +20$ dBmV
$N_2 = -25$ dBmV

Combiner / splitter

$C_2 = +20$ dBmV
$N_2 = -25$ dBmV

$C_2 / N_2 = 45$ dB

$C_2 / N_0 = 42$ dB

Since noise level is constant across the frequency band the output noise No = -28 dBmV + 3 dBmV = -25 dBmV.

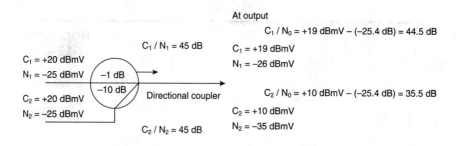

At output

$$C_1 / N_0 = +19 \text{ dBmV} - (-25.4 \text{ dB}) = 44.5 \text{ dB}$$

$C_1 = +19 \text{ dBmV}$

$N_1 = -26 \text{ dBmV}$

$$C_2 / N_0 = +10 \text{ dBmV} - (-25.4 \text{ dB}) = 35.5 \text{ dB}$$

$C_2 = +10 \text{ dBmV}$

$N_2 = -35 \text{ dBmV}$

To combine N_1 and N_2 at the output on a power basis

$-26 = 10 \log n_1$	$\log n_1 = -2.6$	$n_1 = 0.00256$
$-26 = 10 \log n_2$	$\log n_2 = -3.5$	$n_2 = 0.00032$
	$\log n_0 = n_0$	$n_0 = 0.00288$

$$N_0 = 10 \log n_0 = 10 \log 0.00288 = -25.4 \text{ dB}$$

Summary: In the case of the 2-port combiner, carrier-to-noise level for both C_1 and C_2 was decreased by 3 dB. For the case of the −10 dB directional combiner where C_2 is combined with C_1 through the −10 dB port, C_1 only had a ½ dB decrease in carrier-to-noise level, while C_2 had a 9½ dB decrease in carrier-to-noise level.

Cascaded Amplifier Theory

Noise voltage generated by a 75-Ω resistor at 20°C, expressed in dBmV. This voltage appears across the input resistance of the 1st 75-Ω amplifier in a cascade of repeater amplifiers.

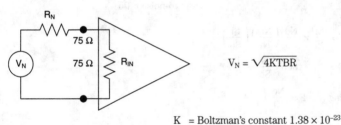

$$V_N = \sqrt{4KTBR}$$

K = Boltzman's constant 1.38×10^{-23}
T = Temperature °K = 20°C + 273°C = 293°K
B = Frequency bandwidth in Hz = 4×10^6 Hz
     for video bandwidth of 4 MHz
R_N = Resistance in ohms

$$V_N = \sqrt{4 \times 293 \times 1.38 \times 10^{-23} \times 4 \times 10^6 \times 75}$$
$$= \sqrt{485208 \times 10^{-17}} = \sqrt{4.85 \times 10^{-12}} = 2.2 \times 10^{-6} \text{ V}$$
$$= 2/\mu\text{V}$$

This is the noise V_N generated by R_N and appears across a circuit of 2 75-Ω resistors connected in series.

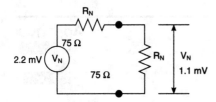

V_{IN} is that portion of noise voltage that appears across the amplifier input terminal and is ½ V_N according to the voltage divider principle.

$$\text{In dBmV} = 20 \log \frac{1.1 \times 10^{-6}\ V}{1 \times 10^{-3}\ V} = 20 \log 1.1 \times 10^{-3}$$

$$= 20 \times -2.959 = -59.2 = -59\ \text{dBmV}.$$

This level in dBmV constitutes the so-called noise floor in cable television amplifier cascades.

Input   Output   Si = Signal input power level
Si   So   Ni = Noise input power level
Ni — G > — No   So = Signal output power level
                   No = Noise output power level
                   G = Gain of amplifier

For the amplifier the noise factor is given as $F = \dfrac{\frac{Si}{Ni}}{\frac{So}{No}} = \dfrac{\text{Input signal-to-noise ratio}}{\text{Output signal-to-noise ratio}}$

Rewriting $F = \dfrac{Si}{Ni} \times \dfrac{No}{So} \left(\dfrac{No}{Ni}\right)\left(\dfrac{Si}{So}\right)$

Since $G = \dfrac{So}{Si}$ then $\dfrac{Si}{So} = \dfrac{1}{G}$

Eq A   substituting $F = \dfrac{No}{Ni}\left(\dfrac{1}{G}\right)$

The noise power output of an amplifier consists of the input noise level amplified G times, plus the noise Nn generated by the amplifier itself.

Eq B   mathematically $No = Ni(G) + Nn$

Eq C   combining Eq A and B $F = \dfrac{Ni(G) + Nn}{Ni(G)} = 1 + \dfrac{Nn}{Ni(G)}$

Consider a cascade of 3 amplifiers 1, 2 and 3

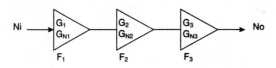

For the cascade $F_C = \dfrac{No}{Ni(Gc)}$ Gc gain of cascade

No noise power output can be calculated

Eq D $No = Ni\,(G_1)\,(G_2)\,(G_3) + Nn_1\,(G_1)\,(G_2) + Nn_2\,(G_3) + Nn_3$

Cascade	Noise generated	Noise generated	3rd amplify.
input noise	by amplifier 1	by amplifier 2	noise

Eq E   also $Gc = (G_1)\,(G_2)\,(G_3)$
For the cascade

Eq F substituting Eqc in A $F = \dfrac{Ni\,(G_1)\,(G_2)\,(G_3) + Nn_1\,(G_2)\,(G_3) + Nn(G_3) + Nn_3}{Ni\,(G_1)\,(G_2)\,(G_3)}$

Rewriting Eq F $F = 1 + \dfrac{Nn_1}{NiG_1} + \dfrac{Nn_2}{Ni(G_1)(G_3)} + \dfrac{Nn_3}{Ni(G_1)(G_2)(G_3)}$

$$\uparrow$$

$$F_1\ (Eqc)$$

and $F_2 = 1 + \dfrac{Nn_2}{NiG_2}$ $F_3 = 1 + \dfrac{Nn_3}{Ni\,G_3}$

Rearranging the above for F_1, F_2, and F_3

$$F_1 - 1 = \frac{Nn_1}{NiGi} \quad \text{and} \quad F_2 - 1 = \frac{Nn_2}{Ni(G_2)} \quad \text{and} \quad F_3 - 1 = \frac{Nn_2}{NiG_3}$$

Substituting back in Eq F.

Eq G for the $F = F_1 + \dfrac{F_2 - 1}{G_1} + \dfrac{F_2 - 1}{G_1\,G_2}$ Eq H F in dB = 10 log F
cascade   F in dB is the noise figure

Therefore Eq G for M stages becomes:

$$F = F_1 + \frac{F_2 - 1}{G_1} + \frac{F_3 - 1}{(G_1)\,(G_2)} + \dots + \frac{F_n - 1}{G_1\,G_2 \dots Gn - 1}$$

Given A 2
Amplifier cascade connected by a piece of cable with its loss equal to an amplifier's gain. The cable will generate noise and has negative gain.

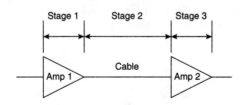

$F_1 = 10$ dB	$F_2 = 10$ dB	$F_3 = 10$ dB
$G_1 = 10$ dB	$G_2 = -10$ dB	$G_3 = 10$ dB

Typically, F & G are given in dB form, therefore to use
Eq G F & G have to be converted to power ratios.

For amplifier 1 $10 = 10 \log F_1, \log F_1 = 1, F_1 = \log^{-1} 1, F_1 = 10$
                         in like manner $G_1 = 10$

For the cable $F_2 = 10$ $-10 = 10 \log G_2$ $\log G_2 = -1$ $G_2 = \log^{-1} (-1)$
                         $G_2 = 0.1$

For amplifier 2 $F_2 = 10$ and $G_2 = 10$ (same as amp 1)

Substituting $F = F_1 + \dfrac{F_2 - 1}{G_1} + \dfrac{F_3 - 1}{G_1 G_2}$

$Fc = 10 + \dfrac{10 - 1}{10} + \dfrac{10 - 1}{(10)(0.1)} = 10 + 0.9 + 9 = 19.9 = 20$

$FcdB = 10 \log Fc = 10 \log 20 = 10(1.3) = 13$ dB

Rule: Doubling number of amplifiers increases noise by 3 dB.

This rule can be applied to a case of many amplifiers (32) in cascade. Since the output signal
level is constant and the buildup of noise is increased by 3 dB every time the number of amplifiers is doubled, the signal-to-noise ratio expressed in dB decreases by 3 dB every time the
number of amplifiers is doubled.

Since the signals carried on an amplifier cascade are television carriers, for a C/N at amplifier
1 output of 60 dB, consider the example below.

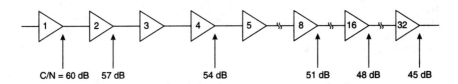

C/N = 60 dB   57 dB   54 dB   51 dB   48 dB   45 dB

The carrier-to-noise ratio is degraded by 3 dB for each time the amplifier number in the cascade is doubled.

45 dB C/N for present-day cable systems is too low. A C/N of 49 dB is acceptable.

To find the C/N of a single amplifier, a test can be performed to find this value. But first consider that for television carrier signal Ci input level & Co output level.

Noise factor $F = \dfrac{Ci}{Ni} \times \dfrac{No}{Co}$

Rewriting in terms of the output carrier to noise level $\dfrac{Co}{No}$

$$\frac{Co}{No} \times F = \frac{Ci}{Ni} \times \frac{No}{Co} \times \frac{Co}{No}$$

$$\frac{CoF}{No} = \frac{Ci}{Ni} \qquad \text{Dividing both sides by F}$$

Eq 1 $\qquad \dfrac{Co}{No} = \dfrac{Ci}{NiF}$

Rewriting Eq 6 in terms of dB.

$$\left.{}^{Co}\!/_{No}\right|_{dB} = Ci_{dB} - (Ni_{dB} + F_{dB})$$

$$= \text{signal input level (dBmV)} - (-59 \text{ dBmV}) - F_{dB}$$

$$\text{C/N}\left|_{\substack{dB \text{ signal amplifier}}}\right. = 59 - F_{dB} + \text{signal input dBmV}$$

The test configuration for the noise figure of an amplifier under test is shown below.

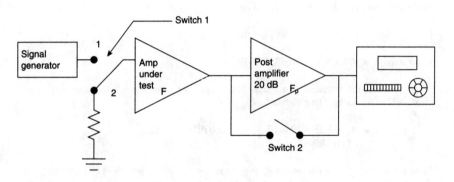

With switch 1 in position 1 and switch 2 closed, gain of amplifier is measured as $\dfrac{Co}{Cin} = \text{gain}$	Signal level meter with noise test feature will not need any correction factor.

1) Or Co/dB – Cin/dB = gain/dB = 22 dB. In terms of dB.

2) Switch 1 is placed in position 2 to terminate amplifier input terminals
   Switch 2 is opened to allow the output noise of amplifier to be further amplified 22 dB by post amplifier. Output noise is read on signal level meter as –20 dBmV.

Actual output noise is calculated by:

Measured output noise = post amp gain + F + input noise + F p + amplifier gain – 20 dBmV = 20 dB + F – 59 dBmV + 22 dB + 7 dB – 20 dBmV = – 59 dBmV + 49 dB + F, F = 10 dB noise contributed.

Now since the noise figure F for a single amplifier has been found, the C/N for a single amplifier can also be calculated.

Example: For an amplifier with a normal carrier input level of +10 dBmV and a noise figure F of 10 dB.

$$Co/No = 59 - F + Cin$$
$$= 59 - 10 + 10 = 59 \text{ dB}$$

Typically CATV amplifiers have noise figures better than 10 dB, usually 4–7 dB.

For the case of identical amplifiers connected together by lengths of cable with a loss equal to each amplifier gain, it was seen that the noise increased by 3 dB every time the number of amplifiers in cascade was double.

$$F \text{ out} = F_1 + 10 \log n$$

F_1 is the noise figure of a single amplifier & N is the number of amplifiers in cascade.

For 2 amplifiers $10 \log 2 = 10(0.301) = 3$ dB.

For 4 amplifiers $10 \log 4 + 10(0.602) = 6$ dB

For 8 amplifiers $10 \log 8 = 10(0.903) = 9$ dB

Using the above mathematical formula the noise figure for a cascade of 7 amplifiers can be calculated. If F_1 for a single amplifier is 5 dB

$$F_{OUT}\Big/_{7 \text{ amplifiers}} = 5 \text{ dB} + 10 \log 7 + 5 \text{ dB} + 8.5 \text{ dB} = 13.5 \text{ dB}.$$

From the example of the cascade of 32 amplifiers, the C/N in dB decreased by 3 dB every time the number of amplifiers in the cascade doubled.

So we can write $C/N\Big/_{\text{cascade}} = C/N\Big/_{\text{single amp}} -10 \log n$

$10 \log n$ is called the cascade factor.

At the 1st amplifier C/N = 60 dB.

Therefore at the 8th amplifier.

$$C/N_{8th} = 60 - 10 \log 8 = 60 - 9 = 51 \text{ dB}.$$

This C/N calculation is important because the signal quality of the amplifier cascade depends on a high signal-to-noise ratio.

If the first amplifier becomes noisy, the whole cascade suffers.

Example: For the 32nd amplifier the C/N has degraded to $50 - 15 = 35$ dB.

This C/N is totally unacceptable

A temporary solution would be to switch the first and last amplifier. Now only service on the end will be affected and not much at that.

In a similar proof, the buildup of amplifier distortion can be shown.

Amplifier distortion appears as second- and third-order distortion.

Second-order distortion involves two frequency components such as the second harmonics of a carrier.

Third-order distortion involves three frequency components. There are three types of third-order distortions.

1) Intermodulation $2 f_1 \pm f_2$ $2f_2 \pm f_1$

2) Cross modulation is caused by any false carrier generated by any third-order distortion with another carrier's modulating signal, which affects any of the correct signal carrier. This is similar to another channel's signal affecting the viewed channel.

3) Third harmonic distortion $3f_1$, $3f_2$ etc

Distortion buildup for a cascade.
For second-order distortion, the cascade calculation is like that for noise.

Second-order distortion for the cascade =
second-order for a single amplifier $- 10 \log n$.

For third-order distortion, i.e., cross modulation, single triple beat (3rd harmonic) or composite triple beat.

*Third-order distortion $\Big/$= third-order distortion $\Big/$ $-20 \log n$
cascade   single
amplifier

*Either type of third-order distortion.

Manufacturers of cable amplifiers specify all of the single amplifier parameters. Therefore the calculations using the formulas can be used to predict the amplifier cascade performance.

Historically the use of push pull type circuitry all but eliminated second-order distortion. Thus third-order distortion became the limiting factor in setting the maximum number of amplifiers in cascade. Noise buildup is still an important specification.

Broadband Signal Test Procedures

Carrier-to-noise ratio is an important parameter that affects signal quality. Early work by the television allocation study organization (TASO) gives the minimum carrier-to-noise ratio specification as 36 dB. This is the minimum specification from the cable TV head-end off-air antenna terminal to the subscriber's television receiver cable input terminals. Cable television systems usually have very noise-free signals at the head-end. The cable system should maintain a carrier-to-noise ratio of 45-dB minimum. With large-screen television sets becoming more commonplace, cable television operators have tightened the specification to 47 dB. Most likely, HDTV will require a 50-dB minimum C/N specification.

This test procedure requires that the carrier level be measured in dBmV and the noise level over 4-MHz bandwidth measured in dBmV. The difference is the C/N measurement.

The connection diagram is shown below.

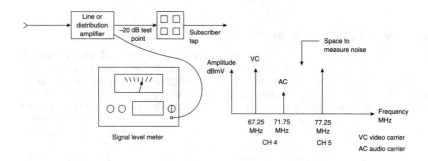

Procedure: Measure carrier level at output test point at CH 5 video carrier. Example +25 dBmV. Now tune below channel 5 video carrier. Remove meter attenuation to read the minimum noise level in dBmV. When minimum level is obtained, lift noise test switch, which increases the meter's bandwidth at the specified 4 MHz. Now read the noise level in dBmV.

Example: –20 dBmV.   The measured C/N = 25 dBmV – (–20 dBmV).
$$C/N = 45 \text{ dBmV.}$$

Low frequency noise and hum testing

Frequencies related to the commercial power system frequency cause a gray-black bar that moves from the bottom of the television to the top at a slow and annoying rate. This is caused by a poor power supply filter in the repeater amplifier's power source. In many cable systems, the 60-Hz ac 60-V power is carried on the cable, which is used by the repeater amplifier's power supply. Power leakage caused by corrosion or bad components can cause this problem. Most present-day signal level meters contain a low-frequency noise and hum test feature.

This test procedure requires an unmodulated video carrier. The connection to the system is usually at the last line extender distribution amplifier test point, as shown below.

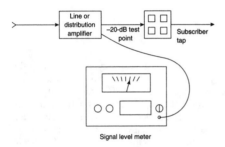

Signal level meter

The unmodulated video carrier is selected by the meter tuning control. Attenuation is adjusted for a reading on the meter of mid to upper scale. The hum test is activated by a selector switch on the meter. The hum level is read on a special meter scale in percent hum. A low-pass filter built into the meter has a cutoff frequency of 120 Hz and is activated by the hum test selector switch. The upper limit of hum is 5%. This test as well as the C/N can be also performed using a spectrum analyzer. This instrument is much more expensive than present-day signal-level meters.

Third-order distortion measurements

Third-order distortion causes beats or false carriers made from the mixing effect of a distorted amplifier or amplifiers. Since all RF carriers on a cable television system are RF carriers modulated by video television signals, the effect of the third-order distortion appears as intermodulation, cross modulation, and composite triple beats. Since cross modulation is the effect of all the system modulation products that interfere with each other's video signals, it is a now-you-see-it/now-you-don't that makes for a difficult thing to measure. Intermodulation is produced by a certain class of false carrier beats $2f_1 \pm 2f_2$ type. Therefore their predicted location in the television spectrum identifies this type of third-order distortion. Composite triple beat examines the buildup of beats. One beat positioned over another builds up to a significant size. With third-order distortion, all three effects will be present. Testing for third-order distortion often means only testing for composite triple beat and intermod. The instrument of choice is the spectrum analyzer and can be used for the three effects.

Intermodulation

Here the spectrum analyzer is connected to any amplifier deep in the system cascade for high enough signal level—usually a line extender amplifier test point or a subscriber tap where the signal level is sufficiently high.

Consider the case of CH3 video carrier at 61.25 MHz (f_1) and channel 3 audio carrier at 65.75 (f_2). Now $2f_1 \pm f_2 = 122.5 \pm 65.75$ MHz for the sum = 188.25 MHz & difference $122.5 - 65.75 = 56.75$ MHz. The sum beat falls between CH 9 audio & video carrier and the difference beat falls between the lower adjacent channel's video & sound carrier (CH 2 video & sound carrier). The connection diagram and the spectrum analyzer display is shown below.

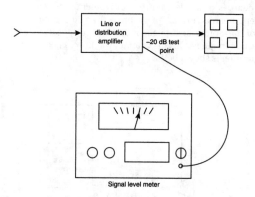

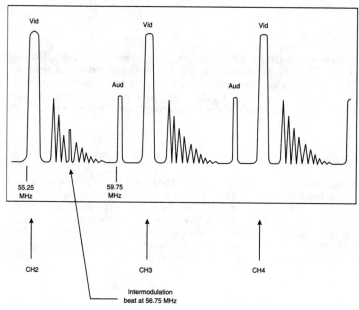

This beat can interfere with channel 2's video signal.

Cross modulation

This test requires the head-end signal to be modulated with the same video signal. If this is not convenient, the head-end source signal can be substituted by a multiple carrier generator with a common horizontal synchronizing signal. The horizontal synchronizing signal is that portion of the video signal with the greatest depth of modulation. Normal television service is interrupted during this test and is often done early in the morning. This system has to be fully loaded for this test, i.e., if it is a 60-channel system, all 60 channels have to be active. For the cross-modulation test, one channel remains unmodulated. Any detected modulation appearing on this carrier has to come from the other carrier's modulation, hence the name cross modulation. A chart or table will have to be obtained for the greatest number of beats falling on beats for, say, the 60-channel system. This chart is available from the instrument manufacturer and/or some of the distribution equipment manufacturers. For example, for a 60-channel 450-MHz cable system, the largest number of third-order beat pileup is 1184 beats appearing near cable channel 33 (277.25-MHz video carrier frequency). This channel will now appear as unmodulated for this test. The test configuration is shown below:

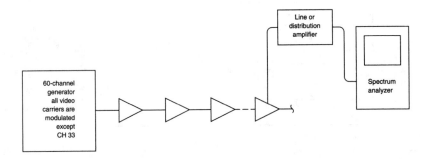

The spectrum analyzer is placed in the zero scan mode, which means it is operating like a receiver tuned to CH 33. The instrument sensitivity is increased and the video modulation appearing on CH 33 is measured. The carrier level to peak modulation is the result and should be greater than 52 dB so as not to be visible. By synchronously modulating all the test carriers except CH 33 makes for a worst-case condition and is easily measured. Still this is a test that requires the system to be shut down.

Composite triple beat

Composite triple beat measures the maximum observed beat level at CH 33 for a 60-channel 450-MHz system. The test connection is the same as for cross-modulation, except all the test carriers have no modulation. The band-pass filter allows the spectrum analyzer to measure the beat buildup level in the space vacated by CH 33. The ratio of this carrier level and the normal test carrier level is the measurement. Example—while CH 33 is turned on, the carrier is measured as +34 dBmV. When this carrier is turned off, the sensitivity of the analyzer is increased and the beat level is measured at –20 dBmV. The carrier-to-beat level is CTB = +34 dBmV – (–20 dBmV) dB = 54 dB.

Coaxial Cable Leakage Testing and Calculations

The leakage specification for cable television systems depends on the frequency band of interest, the leakage intensity in microvolts per meter, and the measuring distance from the cable plant as summarized in the table below:

Frequency Band MHz	Leakage Level Microvolt/Meter	Distance from Plant Feet
Up to and including 54 MHz	15	100
Over 54 MHz, up to and including 216 MHz	20	10
Over 216 MHz	15	100

The test antenna is specified as a dipole. However, a whip antenna mounted on a roving vehicle is often used to locate leaks, which are then measured with the dipole. The detecting receiver is either a signal-level meter of sufficient sensitivity or a special leakage-detecting receiver. Several instrument manufacturers make a combination test transmitter/receiver pair, and cable operators often have several receivers for ongoing leakage monitoring.

Example: If TV channel 6 is used as a test frequency and the leakage measured is at the 20-μV limit, the following formula can then relate this level to a voltage value:

$$E = 0.0207[F(MHz)] [Vr(\mu V)]$$
$$\mu V/M$$

where E = 20 μV/M

$$20 = 0.0207 \,[83.25 \text{ MHz}] \,(Vr(\mu V)$$

$$V_{R(\mu V)} = \frac{20}{0.0207 \,[833.25]} = 11.62 \,\mu V$$

This voltage value in dBmV $= -20 \log \dfrac{1000 \,\mu V}{11.62 \,\mu V} = -20 \log 86.06 = -38.7 \text{ dBmV}$

If the signal level meter cannot accurately measure signals a small calibrated preamplifier can be used.

The dipole antenna should be tuned to the test frequency using the following method:

$$1 \text{ wavelength (ft)} = \frac{984}{f_{(MHz)}} \quad \text{and dipole} = \frac{1 \text{ wavelength}}{2}$$

Each rod of the dipole will be ¼ wavelength.

$$\text{Each rod length in feet} = \frac{984}{4f_{MHz}} = \frac{246}{83.25} = 2.95 \text{ ft}$$

and in inches $2.95 \times 12 = 35.4" = 35½"$

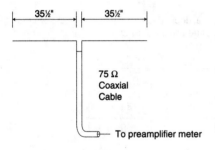

Now that the measured leak level in dBmV can be related to the specified leak in microvolts per meter, a sample calculation of CLI will follow. The method of testing is illustrated below:

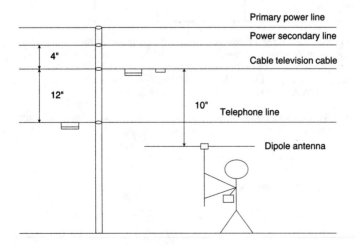

An example of recorded data at TV Ch. 6:

Leak dBmV	Leak mV/m	Leak^2
−39	20	400
−35	30.5	930
−30	54.5	2970
−20	172	29,584
−15	306	93,636
−35	30.5	930
−40	17.2	296
−35	30.6	930
−32	43.2	1866
−30	54.5	2970
	Total	134,512

At a measure level of −30 dBmV corrected by taking any preamplifier gain into account

$$E \text{ mV/m} = 0.0207(83.25)(Vr(\mu V))$$

−30 dBmV corresponds to

$$-30 \text{ dBmV} = 20 \log \frac{V\mu V}{1000 \ \mu V}$$

$$1.5 = \log \frac{V\mu V}{1000 \ \mu V}$$

$$\frac{V\mu V}{1000 \ \mu V} = \log^{-1} - 1.5 = 0.0316$$

$$V\mu V = 31.6 \ \mu V$$

$$EmV/M = (0.0207)(83.25)(31.6)$$
$$= 1.72 \times 31.6 + 54.5 \ \mu V/m.$$

Now: suppose this cable system has 60 miles of total plant and the above data was taken from a 20-mile section. The CLI can now be calculated.

$$CLZ = 10 \log \frac{60}{20} \times 134,512$$

$$= 10 \log [3 \times 134,512] = 10 \log [403,536] = 10 [5.61] = 56.1$$

The upper limit of acceptable CLI = 64. Therefore, this system passes the test.

I

Frequency Testing

Testing for frequency accuracy is necessary for cable television systems so each television carrier has the proper frequency spacing. Errors in channel frequency can affect interaction between adjacent channels, causing the modulation products to mix and in turn affect picture quality. Also, subscriber channel tuning problems will result.

The measurement of frequency was in the past a difficult parameter to measure. With the present-day, highly accurate frequency counter, the measurement process is simplified. Still, presence of video modulation causes the frequency counter to miscount. This problem has to be solved if the measurement is going to be made with the cable service active. Two methods will be studied that will allow modulation of the audio and video carriers.

The first method uses a tunable frequency counter with a built-in signal stripping amplifier. To measure a video RF carrier, the channel is first tuned in by the control on the counter. Since signal level is important so as to not overdrive the counter, an LED bar graph on the instrument indicates proper signal level. Now the instrument controls select the video carrier to be measured, followed by the audio carrier offset test [4.5 MHz]. The connection diagram is shown below:

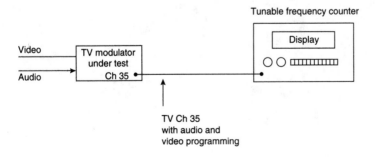

The video carrier and audio carrier offset frequencies are read on the LED numeric display with sufficient accuracy.

Ex. Ch 35 Video carrier   289.2500 MHz
        Offset audio carrier   4.5000 MHz

The second method requires a substitution generator that is tuned to the exact frequency of the modulator to be tested. The exact frequency is determined by zero-beating the two signal sources by a receiver. A signal-level meter can be used as a receiver to indicate zero-beat. The diagram shown below illustrates this procedure:

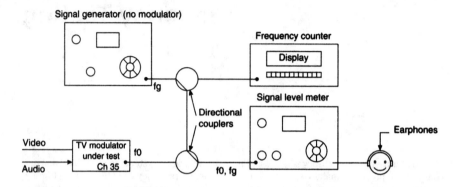

The built-in speaker of the signal-level meter or a pair of earphones can be used to detect zero beat, which means the generator is at the same frequency as the modulator. Directional couplers, and, if needed, an attenuator supply proper signal levels to the signal-level meter and the frequency counter.

For example, to measure the video carrier of Ch 35 @ 289.25 MHz, tune the signal generator to 289.25 as close as the dial allows. Also tune the signal-level meter to 289.25. The signal-level meter sees TV Ch 35 and the generator at 289.25. These carriers should be at approximately the same level. A tone should be heard in the earphones or speaker at the difference between f0 and fg. Use the fine-tuning control on the generator to adjust the tone lower frequency (audio). When the tone in the earphones just disappears, the generator is at the same frequency as the modulator. Now the frequency of the generator is read on the frequency counter.

Analog-to-Digital Conversion

Voice frequency analog voltage conversion to digital code analysis

The conversion process is done in 3 steps.

1) Sampling
2) Quantizing
3) Coding

Sampling

Sampling of a waveform is a sequential measurement of the signal amplitude at fixed even times. This is depicted by a motor-driven switch. When the contacts are closed, the input voltage appears on the output terminals.

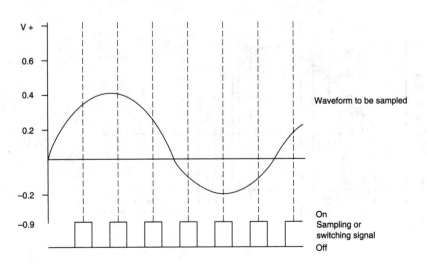

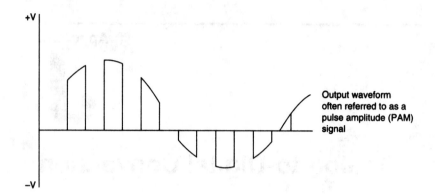

Output waveform often referred to as a pulse amplitude (PAM) signal

Quantizing

This is the process of comparing the pulse amplitude of the PAM signal to a reference signal of a number n discrete levels. This reference level is divided up into a staircase signal with equal voltage levels per step, as shown below:

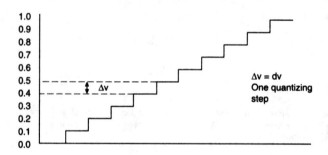

$\Delta v = dv$
One quantizing step

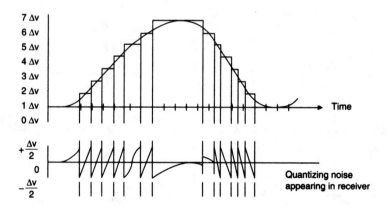

Quantizing noise appearing in receiver

This is a triangular waveform that has an RMS value of

$$V_{RMS} = \frac{V_{MAV}}{\sqrt{3}}$$

when $V_{MAX} = \Delta\frac{v}{2}$ then V_{RMS} (noise) $= \frac{\Delta v}{2\sqrt{3}}$ V

Signal peak – peak amplitude = 2^n dv   Step width × number of steps

For a periodic waveform signal RMS voltage

May be written as $V_{RMS} = \frac{V_{MAX}}{\sqrt{2}}$ $V_{MAX} = \frac{V_{PK-PK}}{2}$

$$\text{so } V_{RMS} = \frac{2^n dv}{2\sqrt{2}}$$

Signal voltage to noise voltage

$$S/N = \frac{V_{RMS} \text{ signal}}{V_{RMS} \text{ noise}} = \frac{\dfrac{2^n dv}{2\sqrt{2}}}{\dfrac{dv}{2\sqrt{3}}} = \frac{2^n dv}{2\sqrt{2}} \times \frac{2\sqrt{3}}{dv}$$

$S/N = \dfrac{2^n\sqrt{3}}{\sqrt{2}}$ and $S/N/dB = 20_n \log2 + 10 \log3$ $- 10 \log2$

$$= 20^n (0.3) + 4.77 - 3.01$$
$$= (6_n + 1.76) \text{ dB.} \qquad \text{Since n = number of steps}$$
$$\text{which = number of bits.}$$

The formula above can be used to calculate the S/N for various numbers of bits.

When n = 7 bits (128 steps or voltage levels).

S/N = 42 + 1.76 + 43.76 dB = 44 dB.

Every time we add one more bit, the S/N improves by 6 dB.

Line-of-Sight Microwave Link

The microwave radio link has been used by the telephone industry to transmit and receive telephone traffic for many years. Early systems used the carrier method of single-sideband suppressed-carrier modulation for transmission of a one-way (half-duplex) telephone call. Another carrier operating on another carrier going the other direction carried the other half of the call. This technique is known as frequency division multiplexing (FDM). Later, telephone voice signals were converted to HRZ digital pulse streams, which is referred to as time division multiplexing (TDM). Both methods were used to stack voice channels on one RF microwave carrier. It is this carrier that is of major concern and should be received at the receiving site in the best possible condition. The basic microwave link for one path is shown below.

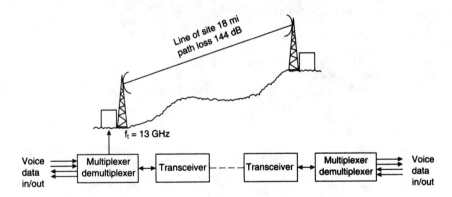

Path length L = 18 miles path loss = 36.6 = 20 log L + 20 log f_t dB

Path loss 36.6 + 20 log 18 + 20 log 13000 = 36.6 + 25.1 + 82.3 = 143.98 = 144 dB

This loss assumes nothing is in the path way and there is no signal refraction.

To examine the effects of refractions, the causes will be investigated. For a line-of-sight (LOS) microwave radio link, objects interfering with the line-of-sight path will refract or diffract the transmitted beam. Since the earth is spherical, the transmitter antenna and receiving towers have to be of sufficient height to clear the earth's bulge. The figure below illustrates this concept.

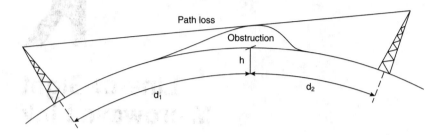

The earth's bulge in feet is given by:

$$h = 0.667 \ d_1 \ d_2 \qquad\qquad d_1, d_2 \text{ in miles}$$
$$h \text{ in feet}$$
$$\text{or } h = 0.079 \ \ d_1 \ \ d_2 \qquad\qquad d_1, d_2 \text{ in km}$$
$$h \text{ in meters}$$

The bulge of the earth has the effect of raising the height of the obstruction. If the earth had no bulge, then of course $h = 0$.

The earth is not purely spherical, and the atmosphere gets thinner with height. This has the effect of changing the earth's radius. The X factor is the ratio of the effective earth's radius to the true earth radius. Mathematically

$$K = \frac{\text{Effective earth radius}}{\text{True earth radius}}$$

K can be negative, infinite, or positive.

When K is – the beam path is concave upward
$K = \infty$ the beam path follows the earth's curvature
For small values of K<1, beam is bent upward
$K = 1$ beam follows optical line of sight.
Temperature also affects K. K is related to h by the expression:

$$h = \frac{0.667 \ d_1 \ d_2}{k}$$

A nominal value of k used is $k = \dfrac{4}{3}$ 1.33

Fresnel zones

The wave front from the transmitting antenna expands as it travels through space, which results in reflections and phase changes as the wave passes over obstacles & obstructions. The height of the transmitting & receiving sites should allow for extra height. The beam cross section shows first, second, and third fresnel zones as concentric bands around the beam center axis, as shown below:

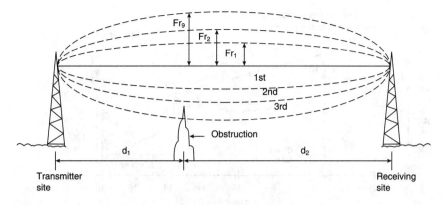

$$Frn = 72.1 \sqrt{\frac{nd_1 d_2}{fD}}$$

d_1 = Distance in miles from transmitter to obstruction

d_2 = Distance in miles from receiver to obstruction

f = Operating frequency in GHz

n = Fresnel zone number

Frn = Fresnel zone radius in feet

$D = d_1 + d_2$ in miles

For metric units d_1, d_2 And D in km
f in GHz and Frn in meters

The formula becomes

$$Frn = 17.3 \sqrt{\frac{nd_1 d_2}{fD}}$$

These formulas are used to adjust tower heights and site elevations so objects or obstructions do not penetrate the 1st fresnel zone.

Path profiling

The procedure of path profiling is to make sure the transmitting and receiving site are clear of obstructions and the path or beam, taking into account the fresnel zones, is far enough above the earth's surface to be unaffected. This procedure might require some on-site work with surveying instruments. However, reasonable estimates can be made using 7.5 min topograhical maps and appropriate graph paper. Also sea-level refractivity contour charts as well as K factor versus refractivity might be helpful in estimating the K factor for the area. A short example illustrates this procedure.

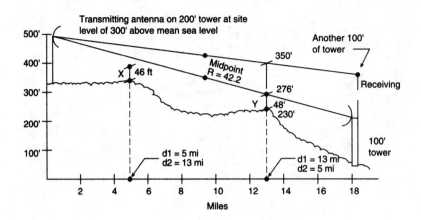

L = 18 mile hop
For the obvious obstruction points X & Y with K = 0.95

Earth's curvature (bulge) at point X $hx = \dfrac{0.667 \times 5 \times 13}{0.95} = 45.6 = 46$

$hy = \dfrac{0.667 \times 13 \times 5}{0.95} = 46$ (Note we are right at beam center at point Y.)

Now the fresnel zone radius must be calculated. This is done for the mid point using the following formula:

$$R = 72.1 \sqrt{\dfrac{25}{13 \times 18}}, \quad R = 72.1 \sqrt{\dfrac{d_1\, d_2}{f_{GHz}\, Lm_1}} = 72.1 \sqrt{\dfrac{81}{13 \times 18}} = 42.4 \text{ ft}$$

To correct this figure for points X and Y, we can use the formula or the graph in the following figure.

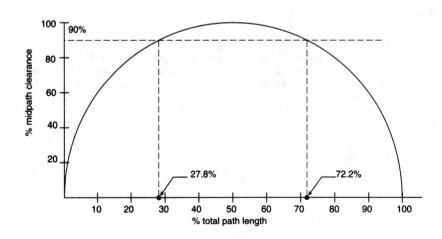

At point X d_1 = 5 mi   Which is $\dfrac{5}{18}$ × 100% = 27.8%

At point Y d_1 = 13 mi   Which is $\dfrac{13}{18}$ × 100% = 72.2%

So at point Y fresnel radius is 0.90 of mid span value of 42.4 ft or 0.90 × 42.4 ft = 38 ft. At this point the bulge of the earth penetrates the fresnel zone and no clearance results.

Another 100 ft of receiving tower height should do the job.
At Y elev. with h = 46' + 230' = 276'
  Corrected fresnel radius = 38

276' + 38 = 314' + path line of site elevation at point Y = 350' so there is clearance allowances for vegetation. Trees might make an additional 50 ft of receiving tower height more prudent.

Once the microwave path is cleared, recall that the 18-mile path loss was 144 dB = PL.

Suppose this was a high-power transmitter with an output power of 1 W converted to dBm. Recall that 1 mW corresponds to 0dBm.

$$P_T = 10 \log \frac{1w}{1 \times 10^{-3w}} = 30 \text{ dBm}$$
$$\underleftarrow{\hspace{2cm}} 1mW$$

If the gain of the transmitting antenna is 45 dB, then the radiated power P_R is calculated.

$$P_R = 30 \text{ dBm} + 45 \text{ dB} = 75 \text{ dBm}$$

At the receiving antenna we may calculate

$$P_{REC} = P_R - P_L = 75 - 144 = -69 \text{ dBm}.$$

If the gain of the receiving antenna is 40 dB, then the receiving power P_{REC} is calculated as

$$P_{REC} = -69 + 40 = -29 \text{ dBm}.$$

If the receiver manufacturer specifies the noise power level as –90 dBm, then the receiver input carrier-to-noise ratio is calculated as:

$$C_R\Big/_{N_r}\Big/_{dB} = C_{R_{dB}} - N_{R_{dB}} = -29 - (-90) = 59 \text{ dB} = 60 \text{ dB}$$

The carrier level is 60 dB greater than the noise level.

A good receiver should be able to produce a baseband S/N of about 52 dB with a 60-dB carrier-to-noise ratio specification.
The fade probability of an outage occuring based on Rayleigh fading the worst month of the year can be calculated by:

$$P_F = T_F \times T_{TF} \times 2.5 \times 10^{-6} \times f \times L^3 \times 10^{-\frac{fm}{10}}$$

P_F = Fade probability of an outage occurring

T_F = Terrain factor   4 for smooth terrain & water

                                    1 for average terrain.

                                    0.25 rocky mounting terrain

T_{TF} = Temperature/humidity factor   0.5 for hot/humid areas
f   = Operating frequency in GHz   0.25 normal climate
L   = Length of path in miles   0.125 high mountain & dry areas
f_m = System fade margin 35 dB AML
                                    30 dB others

Example: Normal terrain/normal climate for the 18-mile hop at 13 GHz and 30-dB fade margin

$$P_f = 1 \times 0.25 \times 2.5 \times 10^{-6} \times 13 \times 18^3 \times 10^{-\frac{30}{10}}$$

$$= 8.1 \times 10^{-6} \times 18^3 \times 10^{-3}$$

$$= 8.1 \times 10^{-6} \times 5832 \times 0.001$$

$$= 47239 \times 10^{-6} \times 0.001$$

$$= 0.047239 \times 10^{-3}$$

For a month consisting of 720 hrs

Amounts to $0.0472 \times 10^{-3} \times 720$ hrs $= 30.5 \times 10^{-3}$ hrs
                                    $= 0.031$ hrs
                                    $= 0.031$ hrs $\times 60$ min/hr
                                    $= 1.86$ minutes in a month

Heavy rain can affect the line-of-sight microwave link. At frequencies below 10 GHz, rain does not have much effect. Studies of rainfall over a five-year period can indicate seasonal rains that could cause problems.

Satellite Ground Station Pointing

The formulas for the earth station azimuth and elevation angles are given as:

Az   Angle in degrees + 180 + $\tan^{-1} \dfrac{\tan D}{(\sin X)}$ from true North

EL   Angle in degrees = $\left[\dfrac{[\cos D \times \cos X] - 0.15126}{\sqrt{[\sin D]^2 [\cos D \times \sin X]^2}} \right]$

X = Earth station latitude in degrees
Y = Earth station longitude in degrees
Z = Satellite longitude in degrees
D = Z − Y

Example: Earth station latitude = 41° 38' 11.0" N
Earth station longitude = 70° 27' 42.0" W
Satellite longitude = 93.5° W

First convert earth station coordinates to decimal degrees.

Latitude		Longitude	
41°	= 41.0000	70°	= 70.0000
38'	= 38/60 = 0.6333	27'	= 27/60 = 0.4500
11.0"	= 11/3600 = 0.0031	42"	= 42/3600 = 0.0117
	41.6364 = 41.63°		70.4617 = 70.46

X = 41.64°   Y = 70.46°   Z = 93.5°   D = 93.5° − 70.46° = 23.04

For the azimuth angle = $180 - \tan^{-1} \dfrac{\tan 93.5}{\sin 41.64}$

$= 180 - \tan^{-1} \dfrac{0.425}{0.66} = 180° - \tan^{-1}(0.644)$

$= 180° + 32.8° = 212.8°$ from true North

For the elevation angle $= \text{Tan}^{-1}\left[\dfrac{[\cos 23.04 + \cos 41.66] - 0.15126}{\sqrt{(\sin 23.04)^2 + (\cos 23.04 \times \text{sim } 46.64)^2}}\right]$

$\qquad = \tan^{-1}\left[\dfrac{(0.920 \times 0.747) - 0.15126}{\sqrt{(0.391)^2\ (0.920 \times 0.664)^2}}\right]$

$\qquad = \tan^{-1}\left[\dfrac{0.6872 - 0.15126}{\sqrt{0.153 + 0.373}}\right]$

$\qquad = \tan^{-1}\left(\dfrac{0.5359}{0.723}\right) = \tan^{-1}\ (0.7389) = 36.5° \text{ from level}$

Since the azimuth angle is from true North, to use a compass in setting the direction, the magnetic difference (declination) must be known. This value can be obtained from a local surveyor or a topographical map of the area.

Example for a magnetic declination of 15.5°

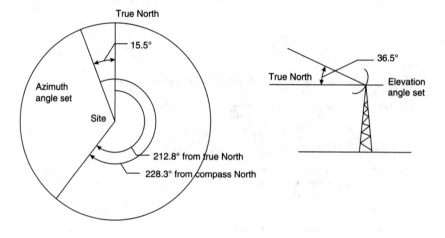

M

MAP/TOP
Broadband LAN Example

Example of a broadband MAP/TOP type local-area network

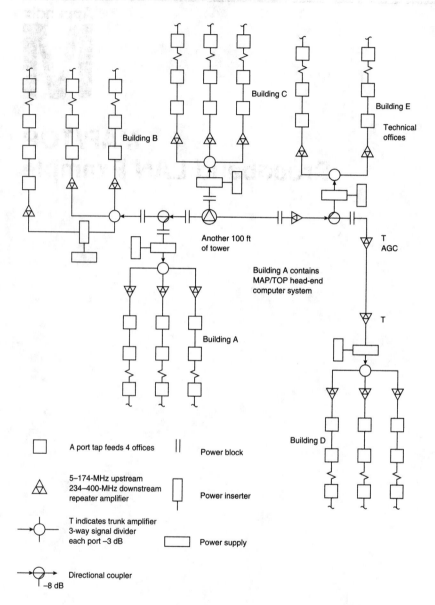

Downstream signals containing modulated carriers that contain digitally encoded manufacturing instructions make up the forward signal. Messages are sent from the head-end to the manufacturing machines and offices.

Upstream signals on modulated RF carriers contain manufacturing data, requests for materials, and schedule progress. These are sent through the head to other offices and warehouses.

The system has nearly equal upstream and downstream frequency bands. Therefore the data channel capacity is equal in both directions.

Tap values have scaled-input port-to-tap port losses to maintain a constant tap signal level as cable loss accumulates. Example below illustrates. The amplifier output is sloped up for a higher level at the upper frequency. Example for the downstream band 234–400 MHz. It is desired to have a tap signal level between +10 dBmV + 15 dBmV.

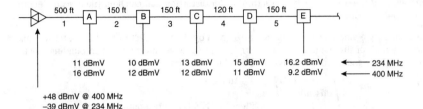

+48 dBmV @ 400 MHz
−39 dBmV @ 234 MHz

System point	Signal levels dBmV	
	@ 234 MHz	400 MHz
At head end	34	48
Cable section	−4	−8
At tap A input	35	40
24-dB tap value		
At tap A port	11	16
Thru loss 24-dB tap	0	−1
Into cable section 2	35	39
Cable section 2 loss	−1	−3
Input to tap B	34	36
24-dB tap value port level	10	12
Thru loss 24-dB tap B	0	−1
Input to cable section 3	34	35
Cable section 3 loss	−1	−3

System point	Signal levels dBmV	
	234 MHz	400 MHz
Input to tap C	33	32
20-dB tap port level	13	12
20-dB tap thru loss	−0.5	−1.5
Into cable section 4	32.5	30.5
Cable section 4 loss	−0.5	−2.5
Into tap 4	32	28.0
17-dB tap port level	15	11
17-dB tap thru loss	−0.8	−1.8
Level in section 5	31.2	26.2
Cable section 5 loss	−1	−3
Input to tap E	30.2	23.2
11-dB tap port	16.2	9.2

As observed in the table, the points that stand out are:

1) The amplifier output is sloped upward.
2) The cable length's loss increases with frequency, thus causing the signal level to be sloped downward.
3) Tap through loss contributes slight downward signal slope.
4) Tap values are chosen to provide signal levels that fall in the specified window of 10 dBmV to 15 dBmV.

Notice: The 1st tap (A) has a level of +16 dBmV @ 400 MHz 1 dB out of the windows. All other taps (B, C, & D) with the exception of E have levels @ 234 MHz and 400 MHz in the window. Tap E with signal levels of +16.2 dBmV @ 234 MHz and +9.2 dBmV @ 400 MHz both values A 1-dB out of the window. If tap E was not the last tap, another amplifier could be inserted which would produce an output identical to the first amplifier.

Usually, design rules: only two distribution amplifiers in cascade are allowed. Trunk amplifiers properly equalized for cable slope can be used to form a cascade to longer & more distant cable runs to outer buildings. Often, trunk amplifiers with automatic gain and slope controls are alternated with manual gain trunk amplifier with thermal condensation only. Cascade lengths (number of amplifiers) are determined by the amplifier carrier noise and carrier-to-distortion specifications. Electrical power is placed on the cable system either at the head-end or, for instance, at each building. Our example shows that each building powers its own cable branch.

Amplifier types are usually of the push/pull integrated circuit type with supportive circuitry. For short cable cascades of up to 10 trunk amplifiers feeding two distribution amplifiers, the normal CATV type amplifiers work well. An estimate of carrier to noise and composite triple beat follows. Composite triple beat is a third-order distortion that can limit the cascade length of push/pull-type amplifiers.

The trunk single amplifier noise figure in dB given by the manufacturer is 62 dB, for a trunk cascade length of three amplifiers. For our example, this assumes unity gain building blocks where amplifier gain numerically equals the cable span loss.

$$C/N \Big|_{Cascade} = C/N \Big|_{\substack{Single \\ amplifier}} - 10 \log N \qquad N = \text{Cascade length (no)}$$

$$C/N \Big|_{Cascade} = 62 - 10 \log 3 = 62 - 4.8 = 57.2 \text{ dB.}$$

3 amplifiers
in cascade

C/N at end of
trunk amplifier

For a single distribution amplifier the C/N is 65 dB.

$$C/N \Big|_{Cascade} = C/N \Big|_{\substack{Single \\ amp}} - 10 \log N \text{ as before}$$

$$C/N \Big|_{Cascade} \quad 62 - 10 \log 2 = 65 - 3 \text{ dB} = 62 \text{ dB}$$

Recall No$dB = 10 \log \dfrac{P1}{P2}$

We now have to combine actual power ratios of the 2 cascades & calculate a new C/N in dB.

So $\log \dfrac{P1}{P2} = \dfrac{No\text{dB}}{10}$

$$\dfrac{P1}{P2} = 10^{\frac{dB}{10}}$$

Base of common
logarithms

Since combining noise powers which is below signal power we may calculate

–57.2 dB trunk 3-A cascade

$$10^{-\frac{57.2}{10}} = 10^{-5.72} = 0.000001905$$

–67 dB distribution 2-A cascade

$$10^{-\frac{62}{10}} = 10^{-62} = \dfrac{0.000000631}{0.000002236}$$

$C/N = -10 \log 0.000002236 = -10 \, (-0.5595) = 56 \text{ dB.}$

In a similar fashion, if the trunk amplifier has single third-order triple-beat specification of 68 dB and the distribution amplifier of 62 dB.

For the trunk cascade

$$\left. C/CTB \middle/ \text{Cascade} \right. = \left. C/CTB \middle/ \begin{array}{l} \text{Single} \\ \text{amplifier} \end{array} \right. -20 \log N.$$

$$= 68 - 20\ (0.477) = 68 - 9.5 = 58\ dB$$

For the distribution amplifier cascade

$$\left. C/CTB \middle/ \text{Cascade} \right. = 62 - 6.02 = 56\ dB.$$

Trunk	-58 dB	$\rightarrow$	$10^{-5.8}$	$= \quad 0.000001585$
Distribution	-56 dB	$\rightarrow$	$10^{5.6}$	$= \quad \dfrac{0.000002512}{0.000004097}$

$$\left. C/CTB \middle/ \text{Cascade} \right. = -10 \log 0.00004097 = 53.88 = 54\ dB$$

For television use the worst case specification is 46 dB and for CTB 52 dB. Therefore this type of LAN will support video transmission very well.

N

Digital Modulation of a Carrier

Most communications methods used today involve digital bit streams containing voice, data, and compressed video. Radio, satellite, and broadband cable systems often use several high-frequency carriers digitally modulated with high-speed bit streams. Where a bit corresponds to only two voltage levels, amplitude modulation methods are not efficient.

For digital signals, the bandwidth efficiency is defined as the number of bits per second per hertz of carrier frequency.

If a bit stream is operated at 100 Mbps and the bandwidth efficiency is 1 bpspHz, then the required bandwidth would simply be 100 MHz. Different modulation schemes have different bandwidth efficiencies.

Modulation methods used for digital signals are:

1) Pulse amplitude modulation (PAM). Two level PAM as shown in Figure A on next page has a bandwidth efficiency of 2 bpspHz.

This form of signals that amplitude-modulate a carrier is susceptible to noise and a distortion causing bit errors. This is measured as the bit error rate (BER).

2) Frequency shift keying, FSK. This signal is also known as two-tone keying. Such a method is diagrammed in Figure B on next page. Each bit is described by either of two frequencies f1 for a binary 1, and f0 for a binary zero.

Early modems used FSK type modulation.

3) Phase shift key (PSK). Each bit corresponds to a separate phase. If for example a binary "1" corresponds to 90° and a binary "0" corresponds to −90° then two-phase states describe 0 and 1 and is known as 2 - PSK. Such a circuit is shown in Figure C on next page.

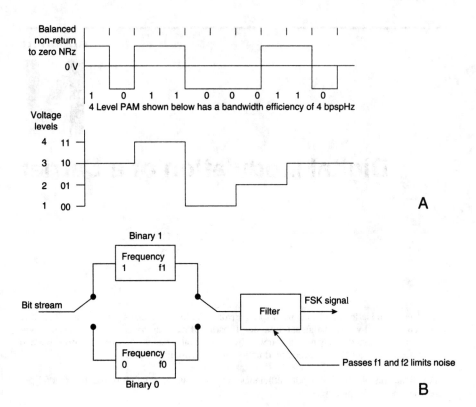

Balanced non-return to zero NRz

0 V

1 0 1 1 0 0 0 1 1 0

4 Level PAM shown below has a bandwidth efficiency of 4 bpspHz

Voltage levels

4	11
3	10
2	01
1	00

A

Binary 1

Frequency
1   f1

Bit stream

Filter   FSK signal

Frequency
0   f0

Binary 0

Passes f1 and f2 limits noise

B

The modulation index for FSK is given as:

$$M = \frac{f_0 - f_1}{fr} \quad \text{if } f_1 = 1000 \text{ Hz}$$
$$f_0 = 1200 \text{ Hz}$$
$$fr = 1000 \text{ Hz}$$

$$M = \frac{1200 - 1000}{1000} = \frac{200}{1000} = 0.2$$

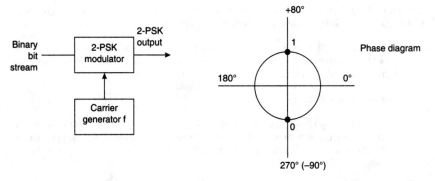

Binary bit stream → 2-PSK modulator → 2-PSK output

Carrier generator f

+80°

1

180°   0°

0

270° (−90°)

Phase diagram

C

If phase states describe bit combinations of 2 bits, this method is known as 4-PSK or QPSK. Every other bit is channeled to a 2-PSK circuit. This method is shown below:

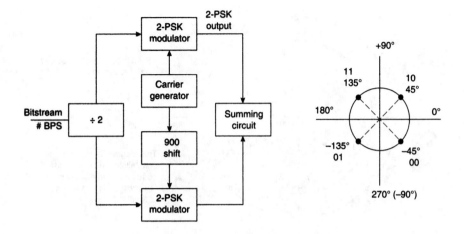

All bits are at 180° or
90° from each other.

Only carrier phase describes the 4 binary bits corresponding to 00, 01, 10, 11.

> 2-PSK has a bandwidth efficiency of 1 bpspHz
> 4-PSK has a bandwidth efficiency of 2 bpspHz
> Extended to 8-PSK has a bandwidth efficiency of 3 bpspHz

A much more efficient method of modulation uses both amplitude and phase changes to describe binary numbers. This is known as quadrature amplitude modulation or QAM.

There are several levels of QAM similar to the PSK level. The bandwidth efficiency of 4 principle methods are given as:

4 QAM	1.5 bpspHz
16 QAM	3.0 bpspHz
64 QAM	4.5 bpspHz
256 QAM	6.0 bpspHz

The diagram for a 16-QAM modulator is shown below. The bit stream is divided in half, every other bit goes through a separate channel I and Q channel.

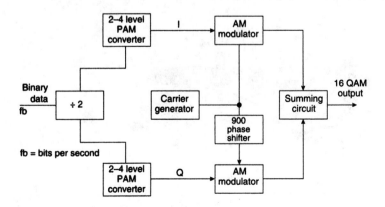

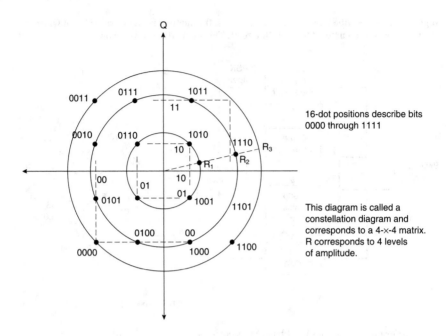

16-dot positions describe bits
0000 through 1111

This diagram is called a
constellation diagram and
corresponds to a 4-x-4 matrix.
R corresponds to 4 levels
of amplitude.

This diagram can be presented on an X - Y oscilloscope by demodulating the I and Q signals where I channel is connected to the horizontal (X axis) and the Q channel is connected to the vertical (Y axis). The oscilloscope's Z or intensity axis should be activated during each bit position.

The oscilloscope constellation diagram should be exactly symmetrical with small clear dots identifying the bits. Smearing dots indicate noise and nonsymmetry phase errors.

Another oscilloscope display that is able to indicate the quality level of QAM systems is termed the eye diagram because the display is similar to an eye. The X axis (horizontal) sweep is adjusted to the symbol or pulse time, and the Y axis (vertical) scale contains signal amplitude information. Digital pulse trains can be analyzed using the eye diagram. A closed eye or fuzzy display indicates noise and signal distortion. The eye diagram is extremely useful when system adjustments are made. Clear eye openings indicate proper adjustment. An eye diagram for a 16-QAM system is shown below.

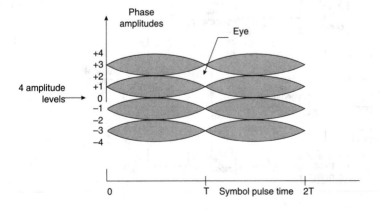

SONET Networks
and Testing Techniques

This appendix was made available through the courtesy of CERJAC, a subsidiary of Hewlett Packard, and is reprinted by permission. The material was chosen because it is extremely well done. The treatment of the subject is straightforward and easy to read, which makes this complicated topic become clear. SONET Network and Testing Technology was developed as a seminar and was presented at various technical conventions and expositions. Therefore there is a continuous progression from the present wired system to the Synchronous Optical Network (SONET). The parallel references to a truck transportation method of delivery is useful in illustrating SONET technology for transporting messages through the optical network.

This appendix provides an excellent overall view of typical network functions and implementations. Signals from various sources arrive at a SONET terminal through ATM, Frame Relay, and other types of networks. The digital data is then packetized according to SONET protocol to make up the payload sections of the synchronous frames. The compatibility of SONET with other types of networks is made clear. Therefore the adding and dropping of data from the SONET system is similar to the on-off ramps of a highway. This appendix provides an excellent, easy way to learn the elements of SONET networks and testing procedures.

Overview of SONET Technology

CERJAC Telecom Operation
43 Nagog Park
Acton, MA 01720

SONET/ATM
Networks and Testing Seminar

Abstract

The Synchronous Optical Network (SONET) is based on a worldwide standard for fiber-optic transmission, modified for North American asynchronous rates. This part of the seminar will cover the technology, concepts and philosophy of SONET. It will cover how SONET works in providing a migration path for existing asynchronous transmission technology. We will discuss the basic SONET technology, and the advantages it offers to both customers and service providers. Some of the issues that arise in deploying a SONET network will also be highlighted.

Author

CERJAC Telecom Operation
43 Nagog Park
Acton, MA 01720

Hewlett-Packard

Overview of SONET Technology

The objective of this seminar is to give a good understanding of SONET transmission technology and how it is being used in telecommunications in the 1990's.

Seminar Content

- Introduction to SONET concepts

- Basic SONET network devices

- Typical network architectures

- The SONET standards

- How SONET works

- Summary

This part of the seminar will cover the technology, concepts and philosophy of SONET. It will cover how SONET works in providing a migration path for existing asynchronous transmission technology. We will discuss the basic SONET technology, and the advantages it offers to both customers and service providers. Some of the issues that arise in deploying a SONET network will also be highlighted.

What is SONET?

Synchronous Optical NETwork

A transmission technology for fiber-optic transmission standardized worldwide which is replacing proprietary point-to-point transmission systems.

The Synchronous Optical Network (SONET) is based on a worldwide standard for fiber-optic transmission, modified for North American asynchronous rates. In the rest of the world, this technology is known as Synchronous Digital Hierarchy (SDH) which is designed to transport CEPT asynchronous transmission rates.

Why SONET?

- Transport system based on worldwide standard
  - Compatible with existing asynchronous equipment
  - Carry multiple tributaries of different types
  - Internetworking between service providers

- Synchronous timing
  - Direct access to tributaries
  - DS3, DS1, DS0

- Includes OAM&P overhead
  - Network management and maintenance support built-in

- Growth for future services
  - ATM

The idea of SONET grew from the concerns of network operators at the lack of standards covering new Fiber-Optic Transmission Systems (FOTS). They saw that these proprietary designs, mainly for DS1 and DS3 transmission, made interworking and signal handover virtually impossible. The T1X1 committee of the Exchange Carriers Standards Association (ECSA) took up the challenge of creating standards.

The SONET standards we have today are the result of the cooperative effort of many interested groups from both private industry and other standards bodies. For example, in 1984 MCI proposed to the Inter-exchange Carrier Compatibility Forum (ICCF) the concept of standardizing the "mid span meet", the handover point of an optical transmission system. This proposal was passed on to ECSA for discussion and elaboration. In 1985 Bellcore proposed the "SONET" concept which was expanded on within the ECSA T1X1 committee. This was submitted to ANSI (American National Standards Institute) for endorsement and released in June 1988 as the ANSI T1.105 standard in use today.

SONET network equipment will also give savings in cost and size. Using a concept called "Direct Synchronous Multiplexing", there is no need to de-multiplex tributary channels before switching, as must be done with the existing DS1/DS3 asynchronous network. By having the optical network operate synchronously, this new concept in multiplexing makes possible circuit re-routing in "real-time".

A further consideration to compatibility is at the network management level. SONET standards include many performance monitoring and reporting systems. By defining the message sets for communicating with Operation Support Systems (OSS), network information may be passed from one network operator to another where desired, and different vendors' NEs will provide the same information to the OSS.

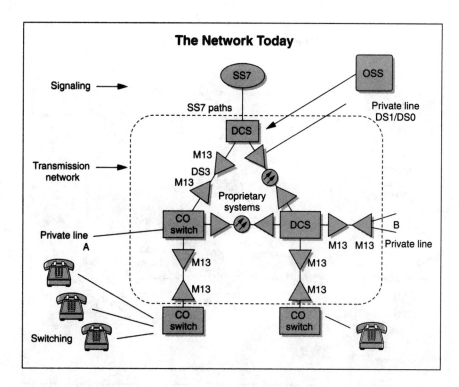

Typically, in existing networks, simple point-to-point transmission technology is used to link the network switches or customer locations. A DS0 signal (64 kbit/s) from a phone call, for example, may be multiplexed up to DS1 using a D4 or D5 channel bank and then to DS3 using an M13 multiplexer. However, to switch the DS0 signal, the full DS3 signal must be demultiplexed. This requires a full set of muxes at each end of a DS3 transmission link. This is also true for higher rate proprietary fiber-optic transmission systems (FOTS). This back-to-back arrangement is very expensive, when in practice only some of the lower-order signals need switching.

Some circuits may use manual "patch panels" rather than Digital Cross-connect Systems (DCS) to route a circuit through the network. Re-provisioning in the event of a customer no longer needing a facility is very time consuming and expensive if re-patching is needed or equipment has to be re-located or retrieved. Even with existing DCS systems, re-routing circuits can take from minutes to hours depending on the control methods.

With a SONET network, all bandwidth allocation and transmission routing can be controlled remotely, making it simple to re-route or re-provision circuits. This can be performed within the SONET overhead, making it easier and less expensive to reconfigure the network "on-the-fly".

The primary issues with today's network are:

- higher rate line systems are proprietary,
- the network architecture is inflexible and expensive for newer services, and
- there is limited built-in network management and maintenance support capabilities

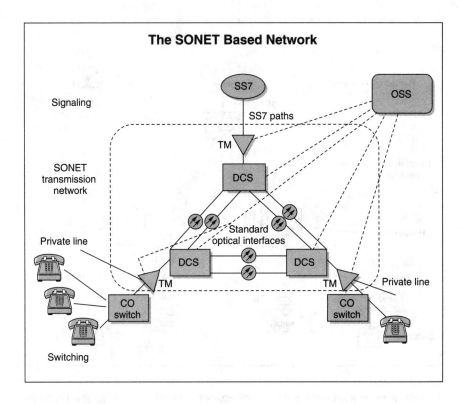

The SONET network has exactly the same function as the existing asynchronous DS3 based network i.e., it will transport customer data from one location to another. However, the SONET standards define transmission rates to 2.4 Gbit/s. This is a massive increase from the existing DS3 rate of 45 Mbit/s.

New bandwidth intensive services like High Definition TV, ultra high-speed computer links, Local Area Network (LAN) bridge links and Broadband ISDN will all be possible. In theory, the SONET network will be viewable as a transmission "cloud". A customer need only be connected to the nearest "Access Device" at the locations between which he requires service.

The OSS provisioning system will then define circuit paths between the locations making use of spare capacity on the SONET transmission systems. With a software-configurable network, in the event of network failure, the routing may be changed within milliseconds.

With this ability to re-route, SONET will take care of equipment failures with negligible effect on customer service.

Comparison of Existing Rates and SONET/SDH

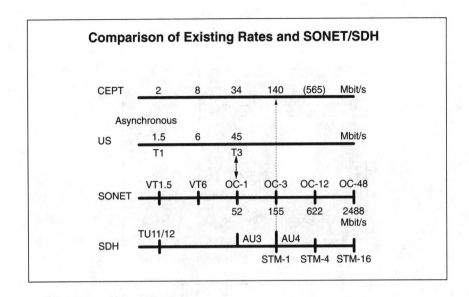

SDH stands for "Synchronous Digital Hierarchy" and is the ITU (CEPT) international standard for SONET transmission.

For SONET to be compatible with existing technology it must be able to carry existing customer services e.g., DS1, DS3. As you can see from the comparison, the SONET line rates have been chosen to match at the lowest level the DS3 customer service. In the CEPT multiplex hierarchy, used outside the USA, a rate of 155 Mbit/s was chosen to match the 140 Mbit/s standard for high capacity bearer transmission.

At present, SONET and SDH standards define signal structures up to 2.4 Gbit/s (OC-48). Higher rates at 4.8 Gbit/s (OC-96) and 9.6 Gbit/s (OC-192) are currently being investigated.

NOTE: The SONET "Optical Carrier" (e.g., OC-3) definitions translate to "Synchronous Transport Module" types (e.g., STM-1) in the SDH standards.

The transmission capability afforded by SONET is massive compared to previous fiber-optic transmission speeds. The SONET standards define FOTS as Optical Carrier (OC) types from OC-1 for existing DS3 service to OC-48 which is a 2.4 Gbit/s FOTS.

SONET Signal Hierarchy and Line Rates

Synchronous transport signal	Line rate Mbit/s	Optical carrier
STS-1	51.84*	OC-1
STS-3	155.52*	OC-3
STS-12	622.08*	OC-12
STS-24	1244.16*	OC-24
STS-48	2488.32*	OC-48

*Denotes the popular SONET physical layer interfaces

The lowest level SONET signal is called the Synchronous Transport Signal level 1 (STS-1) which has a signal rate of 51.84 Mbit/s. The optical equivalent of the STS-1 is the Optical Carrier level 1 signal (OC-1) which is obtained by a direct electrical to optical conversion of the STS-1 signal.

Higher level signals, obtained by byte-interleaved multiplexing lower level signals, are denoted by STS-N and OC-N, where N is an integer. The line rate of the higher level OC-N signal is N times 51.84 Mbit/s, the line rate at the lowest level.

The SONET standard allows only certain values of N. These values of N are: 1, 3, 9, 12, 18, 24, 36, and 48. Values of N greater than 48 may be allowed in future revisions. A maximum value of 255 is allowed under the current standard.

At present, the important SONET line signals and rates are OC-1 at 51.84 Mbit/s, OC-3 at 155.52 Mbit/s, OC-12 at 622.08 Mbit/s, and OC-48 at 2.488 Gbit/s.

Who is using SONET?

- Telephone companies
  - RBOCs, independents

- Long haul carriers
  - AT&T, MCI, Sprint, Wiltel

- Competitive access providers
  - MFS, Teleport

- Private network operators
  - Power companies, Boeing Computer Services

- High-bandwidth end-users
  - VISTA net

A wide range of telecom service providers will benefit from installing SONET technology. From existing telephone companies and private network operators to the operators of high speed city wide MANs, all will eventually move to SONET. The benefits of using SONET are many.

SONET technology is becoming the primary transmission technology employed for telecommunications services; most telecommunications service providers are now installing SONET equipment. Existing phone companies, private network operators, competitive access providers, and operators of high capacity WAN/MANs have embraced SONET. New services, particularly ATM require SONET and are increasing demand further.

All major equipment vendors make SONET transmission and multiplex technology.

Why do they want it?

- Common standard—multi-vendor interoperability

- Better management on OSS

- Fast provisioning

- Better network survivability

- Simpler handover

- Support for future services

- COST

Interoperability: With a common standard, compatible FOTS equipment will be available from many vendors. In a highly competitive market prices will be very attractive. Also, the service provider can easily network with other service providers.

Better network management: With better network management, operators will be able to more efficiently use the network and provide better service. The concept of OSS (Operations Support Systems) is under study by Bellcore. Some OSS standards defining management system interfaces already exist.

Faster provisioning: If new circuits can be software defined to use existing spare bandwidth then provisioning will be much faster. The only new connection needed will be from the customer's premises to the nearest network access node.

Better network survivability: With "real-time" re-routing possible the OSS will be able to take care of failure by simply reprogramming circuit paths. The built-in SONET protection and reporting systems will automatically take care of simple transmission failures.

Simpler handover: If all networks use equipment to the same standard the handover of circuits at the "mid-span meet" should be trouble free.

Support of future services: Looking to the future, the SONET design will cater for new services like High Definition TV, CAD/CAM systems, Wide Area Network backbone networks, Broadband ISDN and new bandwidth-on-demand services. As the SONET operator will have total control of bandwidth allocation, any new service will be simple to provision.

COST: SONET provides a system that is cheaper to install, maintain and provide a given level of service. It will allow the network providers to increase their services while maintaining or lowering their costs.

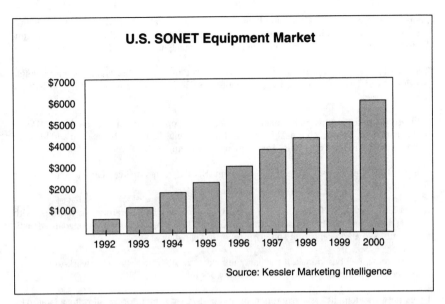

U.S. SONET Equipment Market

Source: Kessler Marketing Intelligence

Network operators are now installing SONET equipment rapidly with many field trials and the appearance of more equipment manufacturers. Most transmission equipment purchased for US networks is now SONET, with new services (e.g. ATM) continuing to drive growth of SONET.

Basic SONET Network Devices

- Digital loop carrier systems
- Terminal multiplexer
- Add drop multiplexer
- SONET DCS (Cross connect)
- SONET regenerator
- ATM switch

Having introduced you to the concept of a SONET network, let's now take a look at the network "building blocks" and how they are configured.

These network elements are now all defined in Bellcore or ANSI standards and provide multiplexing or switching functions. These elements (with the exception of add-drop multiplexers) have the same functions as asynchronous transmission equipment.

Loop Carrier Systems (DLC)–Digital Loop Carriers are specialized SONET back-to-back mux systems providing circuit concentration in the local loop market. The elements used are similar to the Terminal Mux but transmission speed is normally limited to 155 Mbit/s (OC-3).

Terminal Mux–simple multiplexing of SONET and standard DS1/DS3 channels onto a single SONET bearer. Usually described as part of a FOTS/FTS if operating at 622 Mbit/s (OC-12) or above.

Add Drop Mux–a terminal multiplexer with the ability to operate in through mode (ADM) and add channels to, or drop channels from, the through signal. This may be used to add, drop or cross-connect tributary channels. They may operate at any SONET rate.

SONET DCS–full cross connect capability at SONET rates using "Direct Synchronous Switching".

Regenerator–for SONET transmission over about 35 miles, regenerators are required with spacing dependent on the transmission technology. These are not just simple signal reconstituters but have alarm reporting and error reporting capability.

ATM Switch–a cell-based switch that may interface to other switches (Network-to-Network Interface or NNI) or user devices (User-to-Network Interface or UNI).

Every network element has alarm reporting and performance monitoring, allowing a fault to be isolated quickly to the individual transmission section with the problem.

Typical Network Architectures

- Point-to-point

- Multi-point/Linear drop

- Rings

- Bi-directional line switched rings

- Path switched ring

- Interconnected SONET islands

SONET technology is being deployed in new installations and to replace or upgrade existing systems when they reach capacity. Using the basic SONET network devices, a variety of network topologies are being used to provide services to customers. These topologies include:

- point-to-point systems,
- linear multi-point systems,
- rings, and
- interconnected SONET networks

Point-to-point

At the simplest level new point-to-point systems will use SONET Terminal Muxes with the ability to expand to more complex SONET constructions later.

This arrangement is similar to that used with M13 muxes in the present asynchronous network. There are two versions of point-to-point systems based on the SONET network elements described earlier:

1. The Digital Loop Carrier system (DLC). These will be used in the local loop to convey concentrated customer circuits to the CO switch. A remote data terminal (RDT) has essentially the same function but the CO mux may have greater management and fault reporting capability. The RDT must also deal with customer interfaces by signaling.

2. Point-to-Point Terminal Multiplexers. These will provide transmission within the SONET network in a similar manner to the M13 muxes today but at higher data rates.

Multi-point/linear drop

An extension of the point-to-point is the multi-point where additional TMs are placed in the point-to-point path to add traffic at new locations. This is where SONET allows flexibility beyond that available with the existing DS3 systems. As SONET multiplexing is simpler this is likely to be a less expensive solution than available with DS3 based systems.

Rings

The ring is a most important new topology possible with SONET. SONET capability goes beyond that of existing systems with the Add-Drop Mux. With the ability to add and drop channels at any ADM, under centralized management control, a very flexible network architecture is possible. Combine this with the reliability of a bi-directional ring topology and a very robust and readily managed network is possible.

Two types of rings have been used in SONET, the path switched ring, and the bi-directional line switched ring.

The path switched ring provides protection switching at the path level. If a failure occurs, the lower level path signal is switched to the protection fiber. Thus live traffic could be carried on both fibers.

In a line switched ring, all traffic at the SONET line level is switched to a protection fiber during a fault. All traffic is normally carried on one of the fibers in the ring.

Interconnected Networks

As greater amounts of SONET equipment is deployed, the network will become highly interconnected. Rings, linear systems and meshes will all be part of the network in which "islands" of SONET equipment are interconnected. These interconnections will lead to many internetworking issues we will discuss.

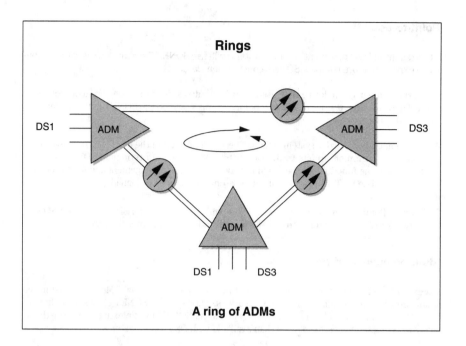

Rings are becoming a key architecture of SONET systems. Should one fiber route be broken, the ADMs can automatically reroute traffic in the opposite direction around the ring. Rings also make it easier to manage services and bandwidth due to the many locations and possible circuit routes to them. Rings are already in service in major metropolitan areas with many new rings being planned and installed.

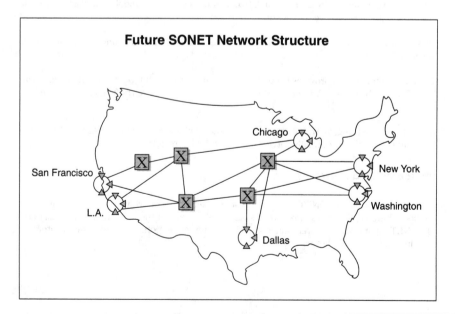

With simplified handover from LATAs to Longhaul Carriers future SONET networks through-out the US may look like this. At each "mid span meet" the interworking of different vendors equipment should be assured if the equipment complies to the standards. However, there will likely be misinterpretations of the standards which will require test equipment to resolve.

Although SONET equipment may be compatible for transmission, it will only be when the management messages have been fully implemented that true interworking will be possible.

The SONET Standards

Bellcore (800)	TR-253	Generic requirements
521 CORE	TR-233	Wideband and broadband DCS
	TR-303 Loop carriers	
	TR-496 ADM	
	TR-499 Transport system	generic requirements
	TR-782 SONET digital	trunk interface
	TR-917 SONET	regenerator equipment
ANSI	T1.105.xx Rates, formats, jitter, etc.	
	T1.106 Optical interface	
	T1.119 OAM&P	communications
	T1.204 Operations,	administration and provision
	T1.231 In-service	performance monitoring

The driving force behind SONET is the standards which define the network equipment functionality and performance. The original Bellcore standard was released in 1985 as TR-TSY-00253. This has been re-issued many times as new requirements are added. GR-TSY-00253 is now in preparation. Other Bellcore standards appeared in November 1989, e.g.. TR-TSY-000496 which defined the operation of the ADM and TR-TSY-000499 (Transport Systems General Requirements).

AT&T and others followed Bellcore in releasing design standards for their own versions of a synchronous network. The SONET submissions have now been amalgamated by ANSI committees into a number of standards, e.g.:

ANSI T1.105 —Digital Hierarchy —Optical Interface Rates and Formats Specification.

ANSI T1.106 —Digital Hierarchy —Optical Interface Specifications.

The T1.105.xx document is now in the process of being updated and revised. ANSI is the co-ordinating/endorsement body for many approved committees like the Exchange Carriers

Standards Association (ECSA) T1X1 committee. The specifications defined by the committees, if approved, are released as ANSI standards.

ANSI (American National Standards Institute) has planned and released the SONET standards in defined Phases. The Phase 1 release covered all hardware operation and interfacing. Phases 2 and 3 covered the management protocols and messages. The reasoning here was that as the protocols are software based it would be simply a matter of software upgrades to bring the already installed SONET equipment up to Phase 3 capability. Manufacturers have the task of upgrading the software to the new standards.

The US SONET standards required some modifications to permit interworking with the networks of other countries. These modifications were incorporated into CCITT standards covering SDH (Synchronous Digital Hierarchy) which is the international equivalent of SONET. The primary standards are CCITT G.707-709.

These standards define the operation of network elements down to the line signal formats, interfaces and security features. It is the in-depth operation of the SONET transmission we will examine next.

How SONET Works

- Overview
  - transmission rates
  - terminology

- The SONET frame structure

- Concept of synchronous frames

In this section we will take a look at the SONET signal from the transmission rates involved to the concepts and structure of the transmitted data. We will describe how an existing asynchronous signal is transported over a SONET network.

SONET Terms

OC-n   Optical carrier n (n = 1 thru 48)

STS-n   Synchronous transport signal n (n = 1 thru 48)

EC-n   Electrical carrier n (n = 1, 3)

SPE   Synchronous payload envelope
       –carrier user data

VT   Virtual tributary
      –subdivision of SPE for lower rates

n = Multiples of basic 51.84 Mbit/s transport signal

SONET, like any other new technology, has many new descriptions, abbreviations and terms. This slide lists the main SONET abbreviations which relate to the SONET transmission concepts:

OC-n —Optical Carrier type "n", where n as we have seen is from 1 (51.84 Mbit/s) to 48 (2.4 Gbit/s). Most popular line signal rates are OC-3 at 155.52 Mbit/s, OC-12 at 622.08 Mbit/s, and OC-48 at 2.488 Gbit/s.

EC-n —Electrical Carrier-n, where n is 1 or 3. This is the electrical interface defined for the 51.84 Mbit/s and 155.52 Mbit/s signal.

STS-n —Synchronous Transport Signal "n". This is the SONET "transport" data structure appearing on the OC-n" FTS. The STS consists of "frames" into which data is filled.

SPE —Synchronous Payload Envelope. A defined area within the STS-N which carries the data for customer services.

VT —Virtual Tributary signal. An SPE can be defined to carry different services. Depending on the data rate of the services the data area within the SPE will be allocated —"Mapped" —to suit. Within the SPE a service will have a defined area, a Virtual Tributary frame.

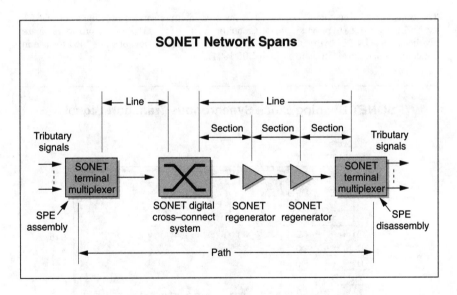

For network management and maintenance purposes, the SONET network may be described in terms of three different network spans:

1. The *PATH* span which allows network performance to be maintained from a customer service end-to-end perspective;

2. The *LINE* span which allows network performance to be maintained between transport nodes and provides the majority of network management reporting; and

3. The *SECTION* span which allows network performance to be maintained between line regenerators or between a line regenerator and a SONET Network Element allowing fault localization.

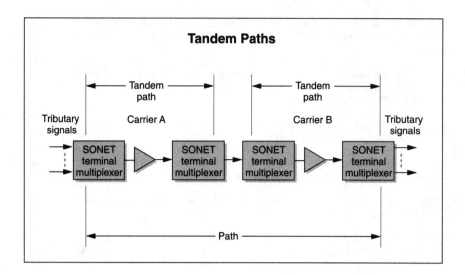

The transport of a customer's signal often involves multiple service providers. As SONET evolved, a need arose to provide a tandem monitor function for use by service providers. Spare overhead capacity (Z5 byte) in the SONET path overhead has now been defined for use in Tandem Connection Maintenance (T1.105.05-1994)

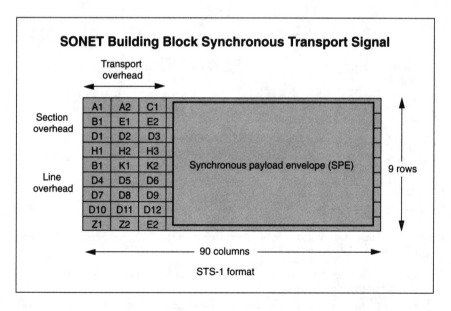

The basic building block of SONET transport is the Synchronous Transport Signal (STS). For clarity, the bits that are transmitted sequentially in the SONET frame are represented in a 2 dimensional matrix of 810 (9 × 90) bytes. 27 of the bytes are the transport overhead, the rest are the Synchronous Payload Envelope (SPE).

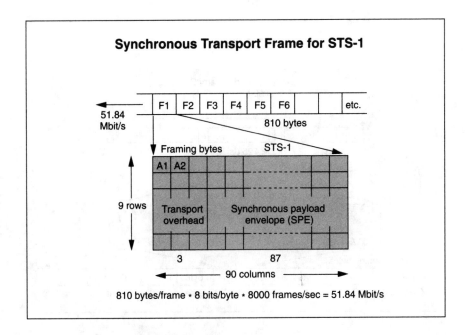

Synchronous Transport Frame for STS-1

810 bytes/frame * 8 bits/byte * 8000 frames/sec = 51.84 Mbit/s

The 2-dimensional map for the STS frame comprises N rows and M columns of boxes which represent individual 8-bit bytes of the synchronous signal. Two framing bytes (A1 and A2) appear in the top left hand box of the 2-dimensional map to provide the frame reference. For an STS-1 frame, there are 9 rows and 90 columns.

The concept of transporting tributary signals intact across a synchronous network has resulted in the term "Synchronous Transport Frame" being applied to such synchronous signal structures. More importantly, however, signal capacity is set aside within a synchronous transport frame to support network transportation capabilities. A synchronous transport frame, therefore, comprises two distinct and readily accessible parts within the frame structure —a synchronous payload envelope part and a transport overhead part.

Synchronous Payload Envelope (SPE): Individual tributary signals (such as a DS3 signal for example) are arranged within the "Synchronous Payload Envelope" which is designed to traverse the network from end-to-end. This signal is assembled and disassembled only once even though it may be transferred from one transport system to another many times on its route through the network.

Transport Overhead (TOH): Some signal capacity is allocated in each transport frame for "Transport Overhead" which provides the facilities (such as alarm monitoring, bit-error monitoring and data communications channels) required to support and maintain the transportation of an SPE between nodes in a synchronous network. Transport overhead pertains only to an individual transport system and is not transferred with the SPE between transport systems.

Scrambling: To ensure that the clock can always be recovered from the received data all but the framing and C1 bytes within the SPE are scrambled.

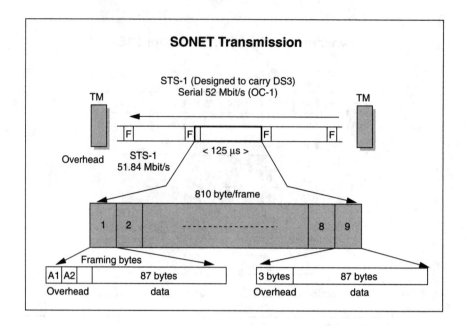

In each frame, the signal bits are transmitted in a sequence starting with those in the top left hand corner byte, followed by those in the 2nd byte in row 1, and so on, until the bits in the 90th (last) byte in row 1 are transmitted. Then, the bits in the 1st byte of row 2 are transmitted, followed by the bits in the 2nd byte of row 2, and so on, until the bits in the 90th byte of the 2nd row are transmitted. The sequence continues with the bytes in the 3rd row followed by the bytes of the 4th row, and so on, until the 90th byte of the 9th row is transmitted. Then the whole sequence repeats.

The synchronous signal comprises a set of 8-bit bytes which are organized into a frame structure. Within this frame structure, the identity of each byte is known and preserved with respect to framing or marker bytes.

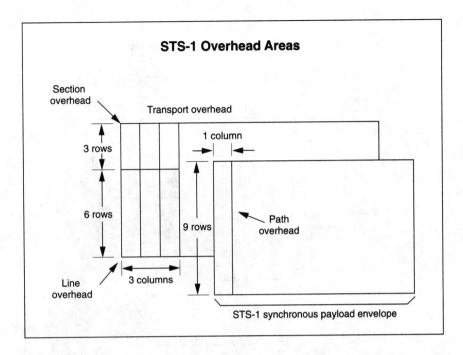

The embedded overhead in the SONET signal is partitioned into three areas in order to support network maintenance at Path, Line and Section levels of network spans.

The Path Overhead provides the facilities required to support and maintain the transportation of the SPE between path terminating locations where the SPE is assembled and disassembled.

The Line and Section Overhead provides facilities to support and maintain the transportation of the SPE between adjacent nodes in the SONET network. These facilities are included within the Transport Overhead part of the transport frame. Thus, the STS-1 Transport Overhead, comprising the first 3 columns of the STS-1 frame, is split between Section Overhead and Line Overhead.

The Section Overhead occupies the top 3 rows of the Transport Overhead for a total of 9 bytes in each STS-1 frame. The Line Overhead occupies the bottom 6 rows of the Transport Overhead for a total of 18 bytes in each STS-1 frame.

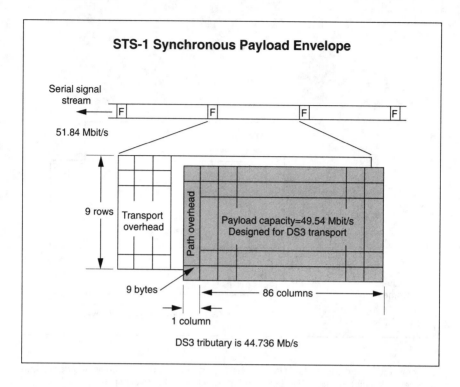

STS-1 Synchronous Payload Envelope

The Synchronous Payload Envelope comprises two parts; a Payload Capacity part and a Path Overhead part.

Payload capacity:

The Payload Capacity area of each SPE is intended to support the transportation of specific tributary signals. The STS-1 payload capacity comprises 774 bytes, structured as 86 columns of 9 bytes. These bytes provide a 49.54 Mbit/s transport capacity with a frame repetition rate of 8000 Hz. This payload capacity has been designed specifically to accommodate the transportation of a DS3 tributary signal (at 44 Mbit/s).

Path overhead:

An area of each SPE is also allocated for Path Overhead. This signal capacity provides the facilities (such as alarm monitoring and performance monitoring) required to support and maintain the transportation of the SPE between end locations (known as Path Termination's) where the SPE is either assembled or disassembled. It also supports the Tandem Connection Maintenance functions. Signal capacity for the STS-1 SPE Path Overhead is allocated in the first column of the STS-1 SPE —a total of 9 bytes per frame.

The Path Overhead comprises 9 bytes and occupies the first column of the Synchronous Payload Envelope. Path Overhead is created and included in the SPE as part of the SPE assembly process. It remains as part of the SPE for as long as the SPE is assembled.

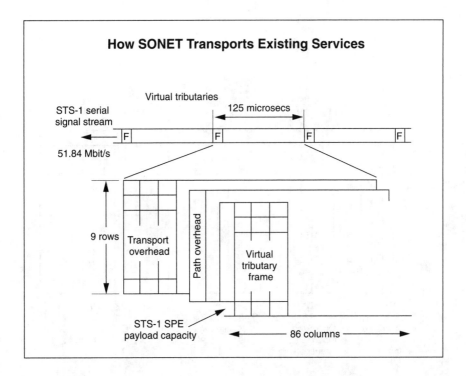

The SONET STS-1 SPE with a channel capacity of 50.11 Mbit/s has been designed specifically to provide transport for a DS3 tributary signal. Transport for a tributary signal with a signal rate lower than that of a DS3, such as a DS1 for example, is provided by a Virtual Tributary (VT) frame structure.

VTs are specifically intended to support the transport and switching of payload capacity which is less than that provided by the STS-1 SPE. By design, the VT frame structure fits neatly into the STS-1 SPE in order to simplify VT multiplexing capabilities. A fixed number of whole VTs may be assembled within the STS-1 SPE.

The concept of a tributary signal (such as a DS3 signal) being assembled into a Synchronous Payload Envelope, to be transported end-to-end across a synchronous network, is fundamental to the SONET standard. This process of assembling the tributary signal into an SPE is referred to as "payload mapping".

To provide uniformity across all SONET transport capabilities, the payload capacity provided for each individual tributary signal is always slightly greater than that required by the tributary signal. Thus, the essence of the mapping process is to synchronize the tributary signal with the payload capacity provided for transport. This is achieved by adding extra stuffing bits to the signal stream as part of the mapping process.

Thus, for example, a DS3 tributary signal at a nominal rate of 44 Mbit/s needs to be synchronized with a payload capacity of 49.54 Mbit/s provided by the STS-1 SPE. Addition of the Path Overhead completes the assembly of the STS-1 SPE and increases the bit rate of the composite signal to 50.11 Mbit/s.

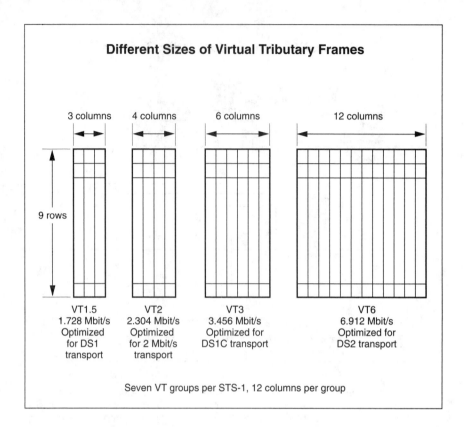

Different Sizes of Virtual Tributary Frames

3 columns	4 columns	6 columns	12 columns
VT1.5	VT2	VT3	VT6
1.728 Mbit/s	2.304 Mbit/s	3.456 Mbit/s	6.912 Mbit/s
Optimized	Optimized	Optimized for	Optimized for
for DS1	for 2 Mbit/s	DS1C transport	DS2 transport
transport	transport		

9 rows

Seven VT groups per STS-1, 12 columns per group

A range of different VT sizes is provided by SONET:

(i)   **VT1.5:** Each VT1.5 frame consists of 27 bytes, structured as 3 columns of 9 bytes. At a frame rate of 8000 Hz, these bytes provide a transport capacity of 1.728 Mbit/s and will accommodate the mapping of a 1.544 Mbit/s DS1 signal. 28 VT1.5s may be multiplexed into the STS-1 SPE.

(ii)   **VT2:** Each VT2 frame consists of 36 bytes, structured as 4 columns of 9 bytes. At a frame rate of 8000 Hz, these bytes provide a transport capacity of 2.304 Mbit/s and will accommodate the mapping of a CEPT 2.048 Mbit/s signal. 21 VT2s may be multiplexed into the STS-1 SPE.

(iii)   **VT3:** Each VT3 frame consists of 54 bytes, structured as 6 columns of 9 bytes. At a frame rate of 8000 Hz, these bytes provide a transport capacity of 3.456 Mbit/s and will accommodate the mapping of a DS1C signal. 14 VT3s may be multiplexed into the STS-1 SPE.

(iv)   **VT6:** Each VT6 frame consists of 108 bytes, structured as 12 columns of 9 bytes. At a frame rate of 8000 Hz, these bytes provide a transport capacity of 6.912 Mbit/s and will accommodate the mapping of a DS2 signal. 7 VT6s may be multiplexed into the STS-1 SPE.

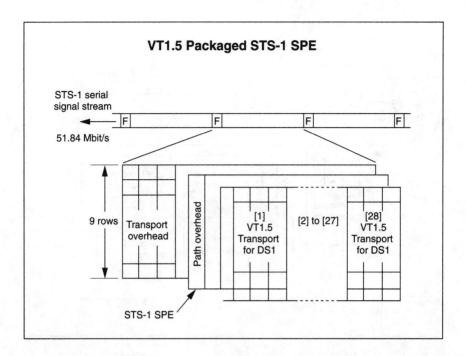

VT1.5 Packaged STS-1 SPE

STS-1 serial signal stream

51.84 Mbit/s

9 rows

Transport overhead

Path overhead

[1] VT1.5 Transport for DS1

[2] to [27]

[28] VT1.5 Transport for DS1

STS-1 SPE

The VT1.5 is a particularly important size of virtual tributary. This is because the VT1.5 is designed to accommodate a DS1 tributary signal which has the highest density of all the tributary signals that appear in the existing network. Twenty-eight VT1.5s can be packaged into the STS-1 SPE for transportation.

The 3 column by 9 row structure of the VT1.5 fits neatly into the same 9 row structure of the STS-1 SPE. Thus, twenty-eight VT1.5s may be packed into the 86 columns of the STS-1 SPE payload capacity. This leaves 2 columns in the STS-1 SPE payload capacity spare. These spare columns are filled with fixed stuff bytes which allows the STS-1 SPE signal structure to be maintained.

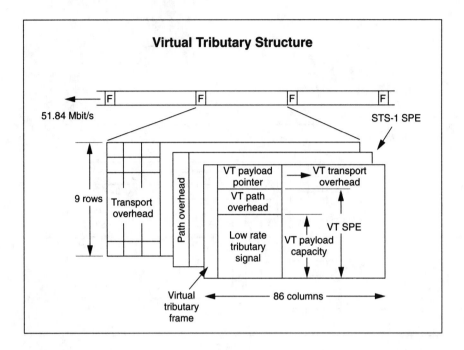

Virtual Tributary Structure

Essentially, the Virtual Tributary Frame represents a mini transport frame structure. It has the attributes of a SONET Transport Frame but is carried within the standard SONET STS-1 Frame structure.

Thus a low rate tributary signal may be mapped into the VT Payload Capacity. VT Path Overhead is added to this payload capacity to complete the VT Synchronous Payload Envelope (VT SPE). The VT SPE is linked to the VT frame by means of a VT Payload Pointer which is the only component of VT Transport Overhead. The VT frame is then multiplexed into a fixed location within the STS-1 SPE.

The VT frame structure is illustrated here as residing in one STS-1 SPE. In fact, however, this frame structure is distributed over four consecutive STS-1 SPE frames. It is, therefore, more accurate to refer to the structure of the VT as a VT multiframe or superframe. The phase of the multiframe is indicated by the H4 byte in the Path Overhead.

Network Functions of the Overhead Capabilities

- Section overhead

- Line overhead

- Path overhead

- VT overhead

The final section of our study of how SONET works considers the embedded overhead capabilities within the signal. The high level of network management possible with SONET depends on the information provided by the "Overhead" areas within the SONET STS-n frames. Typically within the 3 defined overhead areas called Path, Line and Section Overhead (POH, LOH, SOH) there are transmission error detection and reporting features, communication channels, pointers and frame content codes. There is also VT Overhead for VT transport.

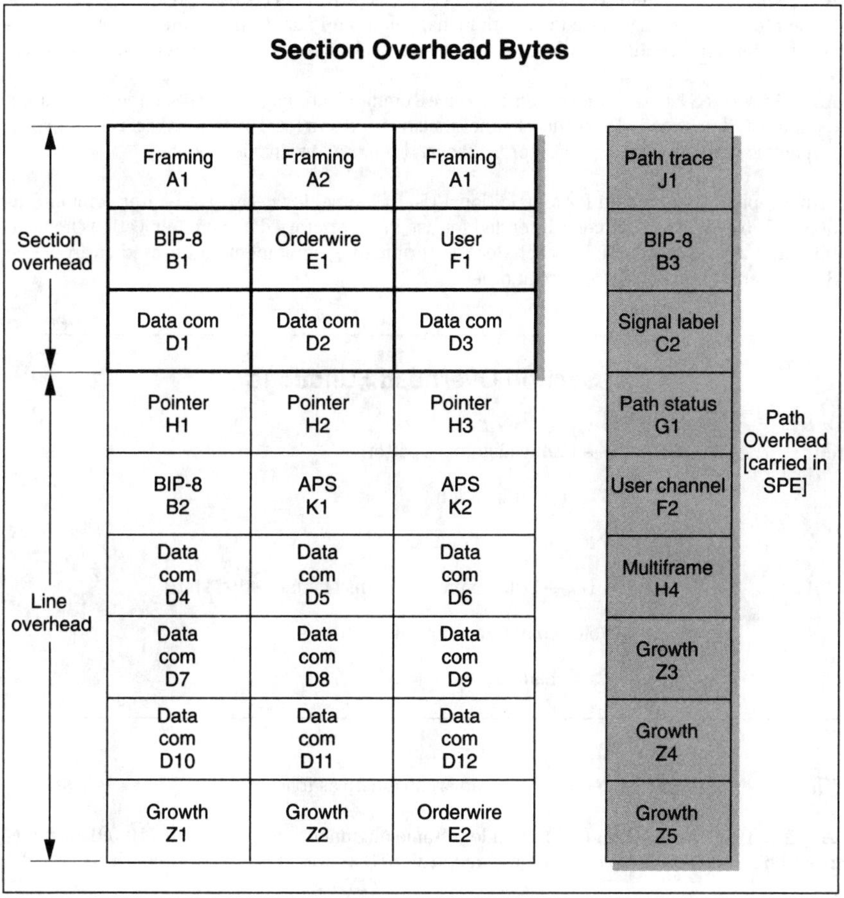

Section Overhead Bytes

Section overhead	Framing A1	Framing A2	Framing A1	Path trace J1
	BIP-8 B1	Orderwire E1	User F1	BIP-8 B3
	Data com D1	Data com D2	Data com D3	Signal label C2
Line overhead	Pointer H1	Pointer H2	Pointer H3	Path status G1
	BIP-8 B2	APS K1	APS K2	User channel F2
	Data com D4	Data com D5	Data com D6	Multiframe H4
	Data com D7	Data com D8	Data com D9	Growth Z3
	Data com D10	Data com D11	Data com D12	Growth Z4
	Growth Z1	Growth Z2	Orderwire E2	Growth Z5

Path Overhead [carried in SPE]

The Section overhead consists of

framing,

ID,

parity,

orderwire,

user, and

data communications channel

Two bytes (A1 and A2) are assigned to the framing pattern in an STS-1 frame for frame alignment. N times 2 bytes, which are byte-interleaved, are assigned to the framing pattern of an STS-N frame. Each STS-1 within an STS-N is identified separately by a binary number corresponding to its order of appearance in the byte-interleaved STS-N frame (C1 byte). An 8-bit wide bit-interleaved parity check (B1 byte) is calculated over all bits of the STS-N frame. The computed value is placed in the Section Overhead of the following STS-N frame.

A 192 kbit/s Data Communications Channel (DCC) is provided (D1–D3 bytes). This DCC is intended to allow message-based network management and maintenance information to be exchanged between Section terminating equipment, such as regenerators and remote terminals.

An orderwire (E1 byte) is provided for voice communications between Section terminating equipment. It is intended that this channel shall be used as a local orderwire reserved for voice communication between regenerators, hubs, and remote terminal locations.

Signal capacity is provided for an additional User Channel for network operator communications (F1 byte). This channel is intended for use in proprietary data communications applications and will be terminated at each Section terminating equipment and passed on from one Section terminating equipment to another.

Section Overhead Functions

- A1, A2 — Frame alignment pattern

- C1 — STS-1 identification

- B1 — Parity check

- D1, D2, D3 — Data communications channel—192 kbit/s

- E1 — Voice communications [Orderwire]

- F1 — User channel

The 9 bytes of the STS-1 section overhead are made up as follows.

A1, A2: Two bytes (A1 and A2) provide a frame alignment pattern (11110110 00101000 binary, F6 28 hex). These bytes are provided in all STS-1s within an STS-N.

C1: The C1 byte is set to a binary number corresponding to its order of appearance in the byte interleaved STS-N frame and can be used in the framing and de-interleaving process to determine the position of other signals. This byte is provided in all STS-1s within an STS-N with the first STS-1 being given the number 1 (00000001).

B1: The B1 byte provides "section" error monitoring by means of a bit-interleaved parity 8 code using even parity (BIP-8). In an STS-N, the section BIP-8 is calculated over all bytes of the previous STS-N frame after scrambling and the computed value is placed in the B1 byte of STS-1 number 1 before scrambling.

E1: The E1 byte provides a local orderwire channel for voice communications between regenerators, hubs and remote terminal locations and is only defined for STS-1 number 1 of an STS-N signal.

F1: The F1 byte is allocated for user's purposes and is terminated at all section level equipment. This byte is defined only for STS-1 number 1 of an TS-N frame. It might be used to download firmware to regenerator.

D1-D3: The three bytes D1, D2 and D3 provide a data communications channel for message-based administration, monitor, alarm, maintenance and other communications needs at 192 kbit/s between section termination equipment. These bytes are defined only for STS-1 of an STS-N frame.

Line Overhead Bytes

Section overhead	Framing A1	Framing A2	STS-1 ID	Path trace J1
	BIP-8 B1	Orderwire E1	User F1	BIP-8 B3
	Data com D1	Data com D2	Data com D3	Signal label C2
Line overhead	Pointer H1	Pointer H2	Pointer H3	Path status G1
	BIP-8 B2	APS K1	APS K2	User channel F2
	Data com D4	Data com D5	Data com D6	Multiframe H4
	Data com D7	Data com D8	Data com D9	Growth Z3
	Data com D10	Data com D11	Data com D12	Growth Z4
	Growth Z1	Growth Z2	Orderwire E2	Growth Z5

Path Overhead [carried in SPE]

The Line Overhead consists of:

- payload pointer
- parity
- APS
- orderwire
- DCC
- growth (unassigned)

The Payload Pointer bytes which provide the linkage between the Transport Overhead and Synchronous Payload Envelope are processed as part of the Line Overhead functionality. They are passed unchanged by section terminating equipment. Separate Payload Pointers are provided for each STS-1 in an STS-N frame. Thus, the Line Overhead provides the necessary support for the networking capabilities of SONET.

The protocol sequences which control Automatic Protection Switching (APS) to alternate stand-by line equipment and plant are signaled as part of the Line Overhead functionality. This functionality provides further support for the SONET networking capabilities. Alarm and parity check information is included as part of the Line Overhead. (These functions are discussed in more detail in the context of in-service testing at the end of this section.)

A 576 kbit/s Data Communications Channel (DCC) allows message-based network management and maintenance information to be exchanged between Line terminating equipment. For example, the routing tables of a digital cross-connect system could be updated by sending the appropriate data over the DCC from the network management computer.

An express orderwire is provided for voice communications between line terminating equipment.

Line Overhead Functions

- H1-H3 — Payload pointers

- K1, K2 — Automatic protection switching

- B2 — Parity check

- K2 — Alarm (line AIS, line FERF)

The 18 bytes of the STS-1 line overhead are made up as follows.

H1-H3: The three bytes H1, H2 and H3 facilitate the operation of the STS-1 payload pointer and are provided for all STS-1s in an STS-N.

B2: The B2 byte provides a BIP-8 "line" error monitoring function. The line BIP-8 is calculated over all bits of the line overhead and payload envelope capacity of the previous STS-1 frame before scrambling and the computed value is placed in the B2 byte before scrambling. This byte is provided for each STS-1 in an STS-N.

K1, K2: The two bytes K1 and K2 provide Automatic Protection Switching (APS) signaling between line terminating equipment and are defined only for STS-1 number 1 in an STS-N. The K2 byte is also used to transmit performance information.

Line Overhead Functions

- D4-D12 — Data communications channel—576 kbit/s

- E2 — Voice communications (Orderwire)

- Z2 — Line FEBE and growth

- Z1 — Synchronization messages

D4-D12: The nine bytes D4 to D12 provide a data communications channel at 576 kbit/s for message-based administration, monitor, maintenance, alarm, and other communications needs between line termination equipment. These bytes are defined only for STS-1 number 1 of an STS-N.

E2: The E2 byte provides an express orderwire channel for voice communications between line terminating equipment and is only defined for STS-1 number 1 of an STS-N signal.

Z2: Part of the Z2 byte is used for line layer Far End Block Error (FEBE), which provides a count of the far end line B2 errors. The remaining bits in the Z2 byte are reserved for future use.

Z1: The Z1 byte has been assigned as a means of passing synchronization messages by the standards committees. The definitions are still in discussion.

Protection Switching

- APS messaging in K1, K2 bytes of first STS-1

- Linear APS

- Ring APS
  –Path switched
  –Line switched
  –Bidirectional line switched—T1.105.01

Protection switching is performed using the messaging capability built into the K1, K2 bytes of the line overhead in the first STS-1 in a SONET signal.

This messaging capability is defined in TR-TSY-00253 and in T1.105. There are different definitions for the messages depending on whether the APS is for Rings or for linear SONET networks. T1.105.01-1994 defines the messaging for bi-directional line switched rings.

Path Overhead Bytes

Section overhead	Framing A1	Framing A2	STS-1 ID C1	Path trace J1	
	BIP-8 B1	Orderwire E1	User F1	BIP-8 B3	
	Data com D1	Data com D2	Data com D3	Signal label C2	
Line overhead	Pointer H1	Pointer H2	Pointer H3	Path status G1	**Path Overhead [carried in SPE]**
	BIP-8 B2	APS K1	APS K2	User channel F2	
	Data com D4	Data com D5	Data com D6	Multiframe H4	
	Data com D7	Data com D8	Data com D9	Growth Z3	
	Data com D10	Data com D11	Data com D12	Growth Z4	
	Growth Z1	Growth Z2	Orderwire E2	Growth Z5	

Path overhead functions provide "end-to-end" monitoring and communications capability. As with the section and line overhead, an 8-bit wide bit-interleaved parity check (B3) is calculated over all bits of the previous SPE. The computed value is placed in the Path Overhead of the following frame.

Alarm and performance information is included as part of the Path Overhead (G1). (These functions are discussed in more detail in the context of in-service testing at the end of this section.)

The structure of the Synchronous Payload Envelope is given via a Path "Signal Label" (C2). This is an 8-bit code value which specifies the SPE structure. 256 different structures are possible. For example, the all "0s" code represents SPE is unequipped (i.e. does not contain any signals).

A fixed length string (64 bytes) is transmitted repeatedly one byte per SPE frame (J1). The 64-byte string can contain any alpha numeric message and is associated with the Path. Continuity of connection to the source of the Path signal can be verified, therefore, at any receiving terminal along the path by simply monitoring this message string.

A generalized multiframe indicator is provided for payloads (H4). At present this facility is used only by VT structured payloads. For example, the VT Overhead and SPE are distributed across 4 frames comprising a VT multiframe. The value of the multiframe indicator in the Path Overhead identifies the phase of the VT multiframe being carried by that SPE.

A User Channel (F2) is provided for proprietary network operator communications between Path Terminating Equipment.

The Tandem connection information (Z5) has recently been adopted to provide Tandem path information for multi-carrier transport of signals.

Path Overhead Functions

- B3—Parity check

- G1—Alarm and performance information
  –FEBE
  –Path yellow

- C2—Signal label
  –Structure of SPE

The nine bytes of STS-1 Path Overhead are carried in the SPE and are made up as follows.

B3: The B3 byte provides a BIP-8 "path" error monitoring function. The path BIP-8 is calculated over all bits of the previous SPE and the computed value is placed in the B3 byte before scrambling.

G1: The G1 byte is used to convey back to the originating STS Path Terminating Equipment the path-terminating status and performance. This feature allows the status and performance of a two-way path to be monitored at either end or at any point along the path.

C2: The C2 byte indicates the construction of the STS SPE by means of a label value assigned from a list of 256 possible values.

Path Overhead Functions

- J1—Path trace
  –Repeating 64 byte message

- H4—Multiframe indication for VTs

- F2—User channel

- Z5—Tandem path performance

J1: The J1 byte is used to repetitively transmit a 64-byte, fixed length string so that continued connection to the source of the path signal can be verified at any receiving terminal along the path.

H4: The H4 byte provides a multiframe phase indication for VT payloads.

F2: The F2 byte is allocated for user's purposes between path termination's, e.g., downloading firmware to terminals.

Z3-Z4: The two bytes Z3 and Z4 are reserved for future use.

Z5: The Z5 byte is used to convey information about the Tandem Connection, including an incoming error count and a tandem connection data link.

H4: The H4 byte provides a multiframe phase indication for VT payloads.

VT Overhead Byte (V5)

- First byte in VT SPE

- BIP-2 (parity)

- VT path FEBE

- VT signal label

- VT path yellow

- Growth

The first byte in the VT SPE is known as the V5 VT overhead byte. This byte provides a:

BIP-2 parity calculation. The parity of the previous VT SPE is calculated across the even bits and odd bits and set in bits 1 and 2

Path FEBE (Far End Block Error). Bit 3 indicates if any errors were received by the path terminating equipment to convey back to the VT PTE the path performance.

Spare capacity for future use. Bit 4 is currently undefined.

VT signal label. Bits 5-7 provide signal label which identifies the VT mappings in use.

Path Yellow. Bit 8 is set to indicate a path yellow. Industry usage is now moving toward calling this failure a Remote Alarm Indicator (RAI)

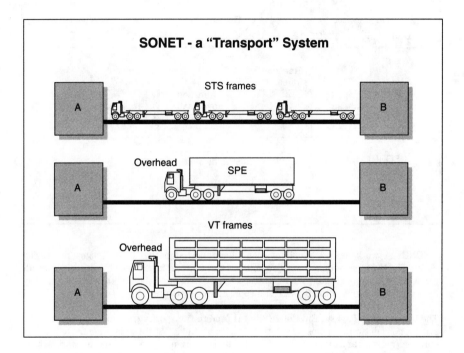

The transport system adopted in SONET is analogous to a road transport system. If you need to deliver items between 2 points you need trucks. Depending on the quantity of items to be moved you need small or large trailers.

Depending on the size of the items being shipped you need "pallets" to allow simple stacking within the trailer payload area. For different item types you have different pallets and different loading instructions. SONET has exactly the same concepts with different names:

• Road —Optical Carrier

• Truck —Synchronous Transport Signal

• Trailer —Synchronous Payload Envelope

• Pallets —Virtual Tributary Frames

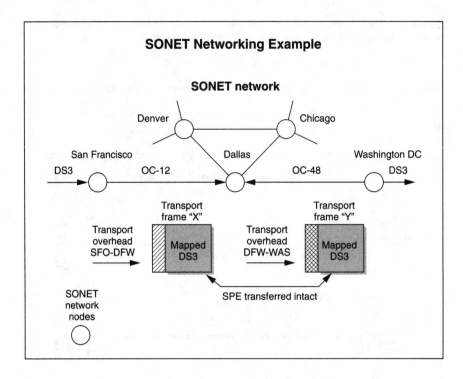

SONET Networking Example

SONET network

A SONET network may be thought of as an interconnected mesh of SONET signal processing nodes. The interconnection of any two nodes in this network is achieved by means of individual SONET Transport Systems. Each transport system carries a signal with a format that may be described in terms of the SONET Transport Frame Structure.

The Synchronous Payload Envelope (SPE) is used to transport a tributary signal across the synchronous network. In most cases, this signal is assembled at the point of entry to the synchronous network and disassembled at the point of exit from the synchronous network. Within the synchronous network, however, the Synchronous Payload Envelope is passed on intact between transport systems on its route through the network.

Transport Overhead is created on the transmit side of each network node and terminated at the downstream receiving network node. Thus, the Transport Overhead pertains only to an individual transport system and supports the transportation of the SPE over that transport system. It is not transferred with the SPE between transport systems.

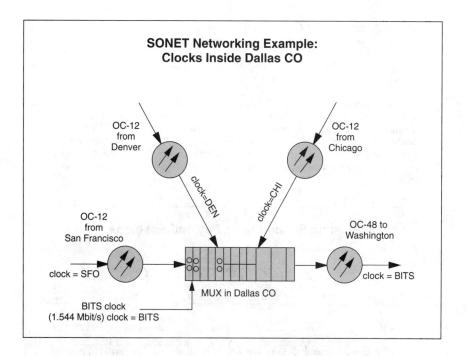

**SONET Networking Example:
Clocks Inside Dallas CO**

At a typical Central Office, SONET signals from different parts of the network may be multiplexed together. The line rates of these may be traceable back to different reference clocks, either because these SPEs were assembled at another phone company using a different clock source, or because of a synchronization fault. The outgoing line rate is defined by the local BITS clock and the small difference between this timing source and the incoming clock needs to be accommodated as discussed next.

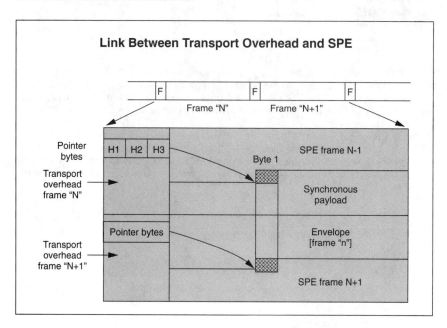

Link Between Transport Overhead and SPE

To take care of small timing differences in the synchronous network, and simplify multiplexing and cross-connection of signals, the SPE is allowed to float within the payload capacity provided by the STS-1 frames. This means that the STS-1 SPE may begin anywhere in the STS-1 payload capacity and is unlikely to be wholly contained in one frame. More likely than not, the STS-1 SPE begins in one frame and ends in the next.

When an SPE is assembled into the Transport Frame, additional bytes, referred to as the "Payload Pointer", are made available in the Transport Overhead. These bytes contain a pointer value which indicates the location of the first byte (byte 1) of the STS-1 SPE. The SPE is allowed to float freely within the space made available for it in the transport frame so that timing phase adjustments can be made as required between the SPE and the transport frame. The payload pointer maintains the accessibility of the SPE by identifying the first byte location of the SPE.

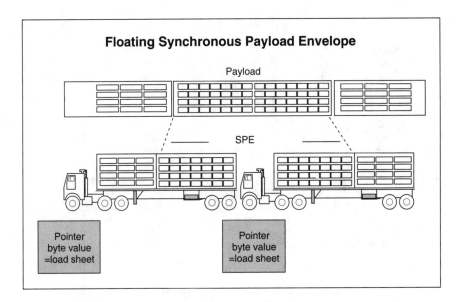

Floating Synchronous Payload Envelope

Payload

SPE

Pointer byte value =load sheet

Pointer byte value =load sheet

This idea of a moving "Payload" (SPE) can be explained using the "trucking" analogy. The truck represents the STS frame and the drivers loading sheet the Transport Over Head (TOH). The payload on the trucks can be split across 2 trucks the split dependent on how much of the load was available at the time the truck had to leave the loading bay.

As SONET is synchronous the trucks are constrained to an arrival and departure schedule. The start position of the load is given by the "load sheet" (TOH) pointer value. Each truck (STS frame) is always fully loaded with "payload" (SPE) of its own and from the truck (STS frame) before. The start position of the new load is given by the pointer values shown on the load sheet (TOH).

Simplified Drop and Insert

- No need to perform stage-by-stage demultiplexing to recover tributaries

- Pointers give exact position of any customer's data

- STS-1 and VT pointers

As the STS-1 pointers give the exact position of any STS payload that payload can be accessed directly without the need to de-multiplex the SONET line signal. This makes devices like the Add Drop Mux and SONET cross-connect switches much simpler than existing telecom systems requiring full demux before switching. In a similar manner any individual customer data channel can be accessed within an SPE.

VT pointers are used to provide the same capabilities and advantages for VT mapped signals, for example, being able to drop a DS1 directly from a STS-1 SPE with VT1.5 mapping.

What Do Payload Pointers Do?

- Allow asynchronous operation within limited clock offsets

- Minimizes network delay compared with existing multiplexers, e.g., M13

- Introduces a new signal impairment –Pointer adjustment jitter

SONET is intended to be a synchronous network. Ideally, this means that all synchronous network nodes should derive their timing signals from a single master network clock.

However, current synchronous network timing scenarios do allow for the existence of more than one master network clock. For example, networks owned by different network operators may have their own independent master timing references. These clocks would operate independently and, therefore, at slightly different rates. Also, the situation in which a network node loses timing reference and operates on a stand-by clock needs to be handled. Synchronous transport must be able, therefore, to operate effectively between network nodes operating asynchronously.

To accommodate clock offsets, the SPE can be moved (justified), positively or negatively one byte at a time, with respect to the transport frame. This is achieved by simply re-calculating or updating the Payload Pointer at each SONET network node. In addition to clock offsets, updating the Payload Pointer will also accommodate any other timing phase adjustments required between the input SONET signals and the timing reference of the SONET node.

Pointers also minimize network delay in the synchronous network. Another approach to over-coming network timing issues is to use 125 μs SPE slip buffers at the inputs to SONET multi-plexing equipment to phase align and slip (repeat or delete a frame of information to correct frequency differences) the individual SONET tributary signals as required. These slip buffers are undesirable, however, because of the signal delay that they impose and the signal impairment that slipping causes. Using Payload Pointers avoids these unwanted network characteristics.

Payload Pointer processing does, however, introduce a new signal impairment known as "Payload Output Jitter." This jitter impairment appears on a received tributary signal after re-covery from an SPE which has been subjected to Payload Pointer changes. Excessive jitter on a tributary signal will influence the operation of the network equipment processing the tribu-tary signal immediately downstream. Great care is, therefore, required in the design of the tim-ing distribution for the synchronous network. This is in order to minimize the number of Payload Pointer adjustments and, therefore, the level of tributary jitter that could be accumu-lated through synchronous transport.

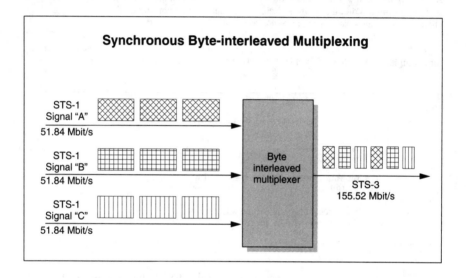

Groups of synchronous transport frames may be packaged for transportation as a higher order synchronous transport signal. Higher order grouping is achieved by means of the process of byte-interleave multiplexing whereby parallel streams of transport signals are mixed together on a fixed byte by byte basis. These parallel streams of transport signals are required to have the same frame structure and bit rate and in addition be frame synchronized with each other.

Thus, for example, 3 parallel and frame synchronized STS-1 SONET signals may be byte-in-terleaved multiplexed together to form an STS-3 SONET signal at 3 × STS-1 bit rate. Byte-in-terleaved multiplexing is accomplished by taking, in turn, one byte from STS-1 signal "A" for output, followed by one byte from STS-1 signal "B" for output, followed in turn by one byte from STS-1 signal "C" for output. At this point, the sequence is repeated by returning to STS-1 signal "A" to provide the next byte for output.

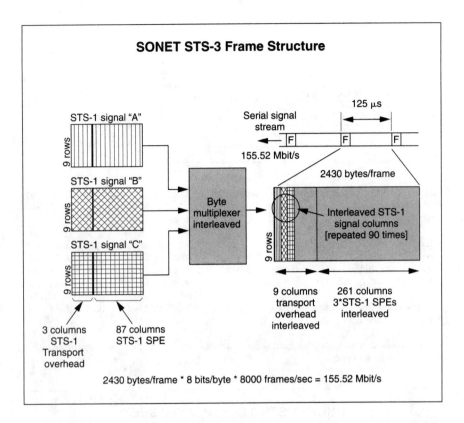

SONET STS-3 Frame Structure

2430 bytes/frame * 8 bits/byte * 8000 frames/sec = 155.52 Mbit/s

The SONET STS-3 signal is assembled by byte-interleaving 3 parallel frame synchronized STS-1 signals. Consequently, a two dimensional map for the STS-3 signal frame comprises the same 9 row depth as the STS-1 signal but has 270 columns which is 3 times the number of columns of the STS-1 signal. The total signal capacity of the STS-3 signal is, therefore, 2430 8-bit bytes or 19,440 bits per frame.

With these frame dimensions and a frame repetition rate of 8000 frames/s, the signal rate for the STS-3 signal is 155.52 Mbit/s. (Note, the frame repetition rate of a SONET signal is 8000 frames/s irrespective of the level of that signal.)

The 2-dimensional map of the STS-3 signal is assembled by taking individual columns from each of the three STS-1 signal structures and interleaving these in a repeating sequence. Thus, starting with the first columns of each STS-1, one column is taken from STS-1 number 1, followed by one column from STS-1 number 2, followed by one column from STS-1 number 3. This sequence is then repeated 90 times until all the columns are assembled.

The first 9 columns of the STS-3 frame are occupied by the Transport Overhead. These 9 columns are not all used completely. The remaining 261 columns are occupied by the three SPE signals associated with the three individual STS-1 signals. These signals are byte-interleaved by columns as described above.

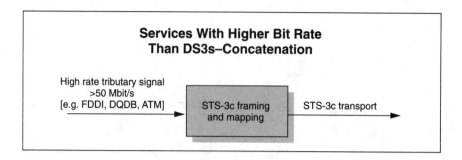

The more advanced customer service signals, such as FDDI, DQDB and ATM (Broadband ISDN), require transport capacity greater than the 49.53 Mbit/s provided by the STS-1 SPE. This is achieved in SONET by means of a higher rate "concatenated" SPE.

A higher rate STS-3 transport signal is normally assembled by byte-interleaving multiplexing three STS-1 transport signals which contain tributary signals at the DS3 signal rate (44.736 Mbit/s) or less. In the SONET context, concatenation means that the higher rate STS-3 transport signal provides effectively one single SPE with a larger payload capacity. A higher rate tributary signal (> 50 Mbit/s) is mapped directly into the larger payload capacity of the STS-3c (where c denotes concatenation) transport signal. The STS-3c SPE is assembled without ever going through the STS-1 signal level.

Once assembled, a concatenated SPE is multiplexed, switched and transported through the network as a single entity.

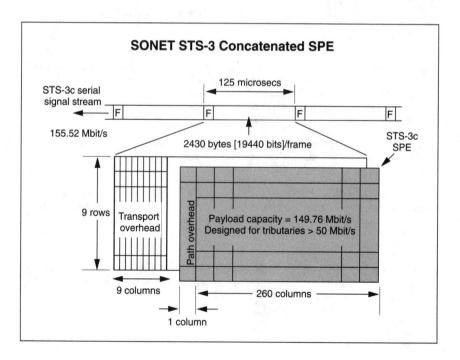

The STS-3c signal frame has the same overall dimensions, 9 rows by 270 columns, the same frame repetition rate, 8000 frames per second, and therefore the same signal rate, 155.52 Mbit/s, as the standard STS-3 signal.

Also in common with the standard STS-3 signal, the first 9 columns of the STS-3c frame, a total of 81 bytes, are allocated to Transport Overhead.

The STS-3c Payload Capacity comprises 260 columns of 9 bytes —a total of 2340 bytes. These bytes provide a transport capacity of 149.76 Mbit/s at a frame repetition rate of 8000 Hz.

Signal capacity for Path Overhead is allocated in the first column of the STS-3c SPE —a total of 9 bytes per frame.

STS-12c is becoming important for supercomputer networking, mapping HPPI and other interfaces into a 600 Mbit/s payload.

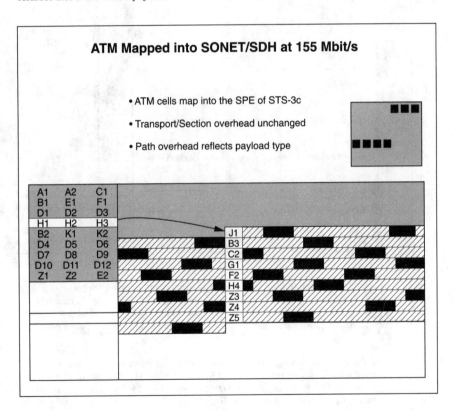

ATM cells (53 bytes, 5 of overhead, 48 of payload) are mapped directly into the SPE without any delineation. This is accomplished by performing a continuous sliding calculation of the Header Error Correction code on 4 preceding bytes.

When the calculation is correct, a cell header may have been detected, and the ATM receiver goes through hunt and sync states to confirm this. The cell stream remains delineated after this until either the path is broken, or serious HEC errors occur. An older proposal to point to cell boundaries with the H4 byte of the path overhead is now obsolete.

Cell scrambling to an $x^{43} + 1$ algorithm is done prior to mapping cells into the SPE. This is in addition to SONET scrambling.

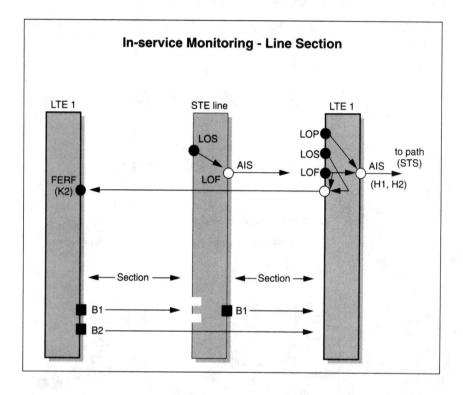

Special signal functions built into particular overhead bytes make effective "in-service" testing possible within a SONET network. Failures such as Loss of Signal (LOS), Loss of Frame (LOF), and Loss of Pointer (LOP) cause Alarm Indication Signal (AIS) to be transmitted downstream. Different AIS signals are generated depending upon which level of the maintenance hierarchy is affected.

In response to the different AIS signals, and failures, other maintenance signals are sent upstream to warn of trouble downstream. Far End Receive Failure (FERF) is sent upstream in the Line Overhead after Line AIS, or LOS, or LOF has been detected by equipment terminating in a line span; a Remote Alarm Indication (RAI) for an STS Path, which is also referred to as STS Path Yellow, is sent upstream after STS Path AIS or STS LOP has been detected by equipment terminating an STS Path.

Similarly, a Remote Alarm Indication (RAI) for a VT Path, which is also referred to as VT Path Yellow, is sent upstream after VT Path AIS or VT LOP has been detected by equipment terminating a VT Path. The path terminating equipment's report to the management OSS the alarm condition.

Performance monitoring at each level in the maintenance hierarchy is based on Bit-Interleaved-Parity (BIP) checks calculated on a frame by frame basis. These BIP checks are inserted in the overhead associated with the Section, Line and Path maintenance spans. In addition, equipment terminating STS Path and VT Path spans produce Far End Block Error (FEBE) signals based on errors detected in the STS Path and VT Path BIPs respectively. The FEBE signals are sent upstream to the originating end of a Path.

Summary

SONET technology

- STS-1

- SPE mapping

- Payload mapping

- Byte interleaving,
  concatenation

Benefits to the SONET user

- Vendor compatibility

- Fast provisioning

- Bandwidth management

- Network survivability

- Potential for new services

Alarms and fault reporting, Control

- Path, Line, Section Overheads

- User channel, DCC

Let's now review the ideas and concepts covered regarding SONET.

What is SONET defined to be?
-a standard for a high-speed optical-fiber-based network.

What devices have been defined as network elements?
-DLC, TM, ADM, DCS and regenerators.

Who has defined the SONET standards?
-ANSI, Bellcore, CCITT.

The theory and operation of the SONET technology?
-OC-n, STS-n, SPE, VTs, Superframes, Concatenation.

The services to be transported by SONET?
-DS0, DS1, DS3, HDTV, BISDN, ATM.

The resilience of the SONET network due to "overhead" providing status, alarm and management information?
-POH, LOH, SOH, management data communication channels, standardized maintenance messages.

The benefits to network operators in changing their transmission networks to SONET?
-commodity network element market due to standards.

Faster and better network management and control?
-provisioning, service re-routing, billing, bandwidth utilization, fault isolation.

Potential for new bandwidth hungry services?
-LAN, MAN, WAN interconnects,
BISDN, CAD/CAM, Video on demand.

Testing SONET Networks

CERJAC Telecom Operation
43 Nagog Park
Acton, MA 01720

SONET/ATM
Networks and Testing Seminar

Abstract

SONET has moved from the R&D labs into the field, bringing with it a host of new installation and maintenance test requirements. The SONET standard allows network equipment from different vendors to be connected at an optical interface point. This compatibility at high rate interfaces enables greater flexibility in configuration of complex networks incorporating rings, linear add-drop systems, and broadband digital cross-connect systems. Testing these new network topologies requires test equipment which can interface at the SONET standard rates.

As SONET deployment continues, new test and maintenance issues will continue to emerge. SONET networks of the future will feature multiple rings, interfaces among SONET network elements at many different OC-N rates, and complex tributary mapping and routing plans. The ability to routinely access and test at these OC-N interfaces, and to map from any one interface rate to any other will be necessary for network implementers and maintenance personnel. And as SONET network complexity increases, so will the requirements for sophisticated test and diagnostic procedures.

This paper discusses situations which can arise in the deployment of SONET technology, and presents test methods to address these new challenges.

Author

CERJAC Telecom Operation
43 Nagog Park
Acton, MA 01720

Hewlett-Packard

Testing SONET Networks

SONET has moved from the R&D labs into the field, bringing with it a host of new installation and maintenance test requirements. This paper discusses situations which can arise in the deployment of SONET technology, and presents test methods to address these new challenges.

Seminar Content

- SONET testing background

- Out-of-service equipment tests

- Network turn-up and service verification

- Alarm and APS testing

- Testing Rings

- Synchronization tests

- Summary

Background to Asynchronous Testing

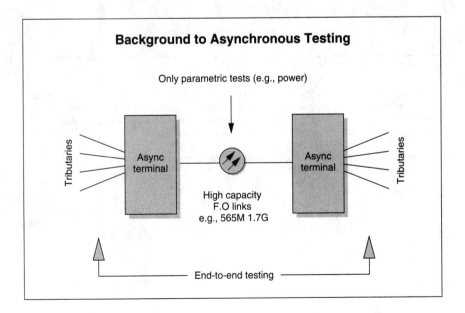

Prior to the deployment of SONET, high speed communication was accomplished using asynchronous fiber-optic transmission systems. Since there was no standard for the optical interface, these systems employed proprietary optical rates and formats.

Testing of these asynchronous systems is carried out end-to-end on a complete line system, usually at a DS3 or DS1 tributary interface. Each tributary channel of the line system is stimulated on one side by means of a test signal incorporating a Pseudo Random Binary Sequence (PRBS). On the other side, errors in the PRBS bit pattern are detected and a Bit-Error Ratio (BER) measurement is made as a measure of performance.

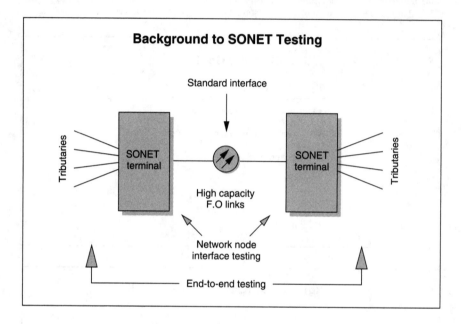

Now, however, with defined standards for SONET, network equipment from different vendors can be connected at the optical interface point. This compatibility at high rate interfaces also enables greater flexibility in configuration of complex networks incorporating rings, linear add-drop systems, and broadband digital cross-connect systems.

Testing these new network topologies requires SONET-compatible test equipment which can interface at the SONET standard rates.

Installation and Maintenance Testing

- SONET equipment is now routinely installed for network growth.

- Services delivered over SONET include DS0, DS1, DS3, OC-3c and ATM.

- SONET interfaces are provided for end users and at inter-network POPs.

- Installation and maintenance of SONET networks is now a high priority revenue-critical activity.

Virtually all major carriers now deploy SONET equipment for network growth. In addition, new competitive access carriers have typically installed all-SONET networks, bypassing the asynchronous stage completely.

With this increasing SONET deployment has come an increased need for installation and maintenance equipment and procedures.

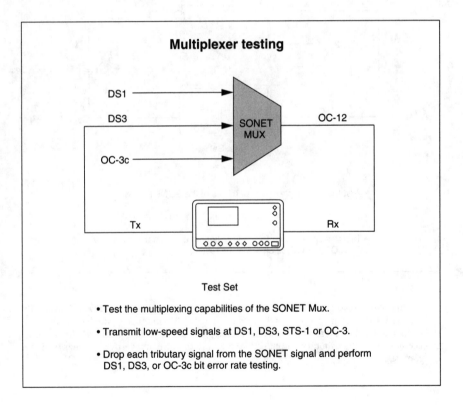

Multiplexer testing

DS1

DS3   SONET   OC-12
                        MUX

OC-3c

Tx   Rx

Test Set

- Test the multiplexing capabilities of the SONET Mux.

- Transmit low-speed signals at DS1, DS3, STS-1 or OC-3.

- Drop each tributary signal from the SONET signal and perform DS1, DS3, or OC-3c bit error rate testing.

Pre-installation testing is sometimes performed on network elements. For example, the performance of the payload mapping process in a SONET multiplexer can be checked.

The SONET tester provides a test stimulus, such as a DS3, DS1, or SONET STS-1, OC-3, or OC-3c signal, normally containing a PRBS running at the desired bit rate. After the test tributary signal has been mapped into a SONET signal by the multiplexer, the SONET tester recovers the test tributary for analysis.

The mapping process is verified by proving the integrity of the recovered tributary signal with a BER test. This requires that a tributary demapping capability must be provided in the SONET tester.

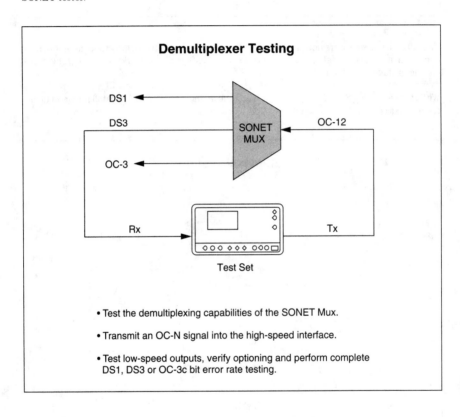

Demultiplexer Testing

• Test the demultiplexing capabilities of the SONET Mux.

• Transmit an OC-N signal into the high-speed interface.

• Test low-speed outputs, verify optioning and perform complete DS1, DS3 or OC-3c bit error rate testing.

For demultiplexer testing, the SONET tester maps a tributary signal into a SONET structure and applies it to the high-speed SONET interface. The SONET multiplexer under test demaps the tributary signal and makes it available at an appropriate tributary interface. The tester is then connected to the tributary output port to measure the BER and verify the integrity of the tributary signal.

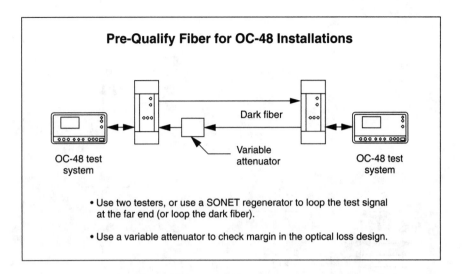

Pre-Qualify Fiber for OC-48 Installations

OC-48 test system

Dark fiber

Variable attenuator

OC-48 test system

• Use two testers, or use a SONET regenerator to loop the test signal at the far end (or loop the dark fiber).

• Use a variable attenuator to check margin in the optical loss design.

Before SONET transport systems are turned-up, it is often necessary to pre-qualify the fiber facilities. This is particularly true when existing fiber is being re-used to support higher rates than originally deployed.

To pre-qualify fiber spans for OC-48 transmission, for example, a test signal can be applied at one end of the dark fiber, and measured by a second test set at the far end. Extended tests can be run, monitoring payload bit errors and SONET overhead parity errors.

It is also possible to employ a SONET regenerator to loop the far-end signal so that a single OC-48 test system can be used. In relatively short distance installations, it may also be reasonable to provide a direct loop of the dark fiber without regeneration.

After confirmation of error-free transmission at nominal design levels, additional loss can be introduced into the optical path using a variable attenuator. In this way it is possible to determine how much margin exists in the optical loss design.

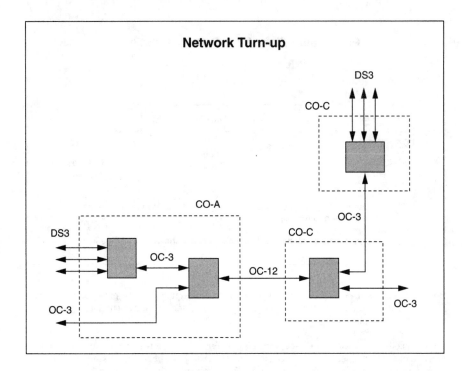

SONET terminals often include diagnostics to facilitate the installation of simple point-to-point configurations. More complex networks, however, are best tested segment by segment before final end-to-end testing is attempted. In addition, as shown above, it may be necessary to turn up and verify each central office separately before connecting the fiber spans.

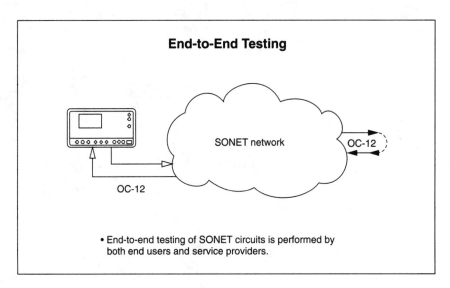

The aim of end-to-end testing is to verify that the integrity of an STS or VT payload is maintained through the network under test.

On the transmit side of the tester, a PRBS is loaded into the selected SPE in a structured SONET signal. This test stimulus is transmitted through the network. At the receive side of the tester, the test SPE is demultiplexed and its integrity verified by a BER measurement on the recovered PRBS.

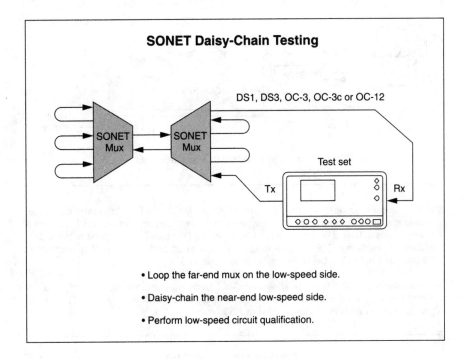

SONET Daisy-Chain Testing

DS1, DS3, OC-3, OC-3c or OC-12

SONET Mux

SONET Mux

Test set

Tx   Rx

• Loop the far-end mux on the low-speed side.

• Daisy-chain the near-end low-speed side.

• Perform low-speed circuit qualification.

End-to-end testing of tributaries can be extended to multiple channels of a transport system through daisy-chaining. This can be done with tributaries at any rate, from DS1 to OC-12, as long as all the tributary channels are at the same rate.

As shown above, the far end tributary ports are each looped back, while a daisy chain connection at the near end directs a single test signal back and forth through the system, simultaneously exercising all the tributary channels.

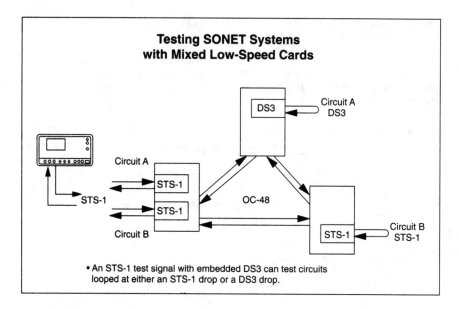

Testing SONET Systems with Mixed Low-Speed Cards

• An STS-1 test signal with embedded DS3 can test circuits looped at either an STS-1 drop or a DS3 drop.

OC-48 transport systems frequently employ either DS3 or STS-1 tributary interfaces. Increasingly, network topologies are being designed which require individual tributaries on OC-48 systems to have STS-1 interfaces on one end and DS3 interfaces on the other end. This facilitates connection to DS3 distribution points on one end, and extension of the circuit via STS-1 to other SONET network elements on the other end.

In these cases, a SONET tester providing an STS-1 signal with an embedded DS3 payload can be used to test either circuit type. The test signal is applied at an STS-1 interface. The far-end signal is then looped back at either STS-1 or DS3.

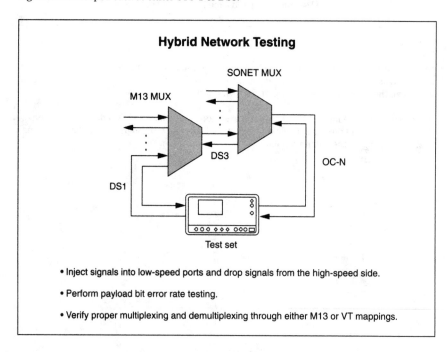

Hybrid Network Testing

• Inject signals into low-speed ports and drop signals from the high-speed side.

• Perform payload bit error rate testing.

• Verify proper multiplexing and demultiplexing through either M13 or VT mappings.

There is an extensive embedded base of DS1/DS2/DS3 (M13) multiplexers in the network. And, although VT1.5 (Virtual Tributary, 1.5 Mbit/s) based equipment is now being deployed, a large number of DS1s will continue to be transported within DS3s over the SONET network.

To test these hybrid network configurations, it is necessary to generate and measure hybrid T-carrier and SONET signals. This requires the ability to inject signals at a low-speed port (DS1 or DS3) and measure the resulting transmission and alarm performance in the multiplexed outputs at high speed SONET ports, or vice versa. Verification of proper multiplexing and de-multiplexing of DS1 signals through either M13 or Virtual Tributary mapping will be a routine maintenance requirement as SONET islands grow.

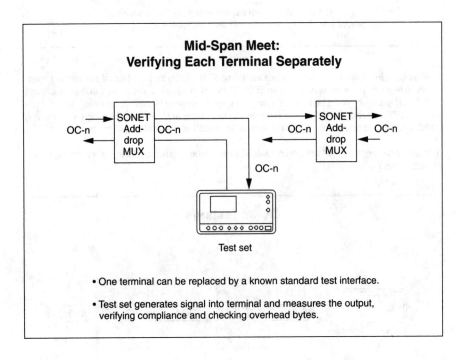

Mid-Span Meet:
Verifying Each Terminal Separately

Test set

- One terminal can be replaced by a known standard test interface.

- Test set generates signal into terminal and measures the output, verifying compliance and checking overhead bytes.

Fiber-optic connection of equipment from different vendors is known as a mid-span meet. Transport mid-span meet, in which basic signal transmission from one terminal to another is achieved, is already a reality among SONET terminals from a growing number of vendors. In the future, mid-span meets will expand to include overhead communications capabilities, in addition to transport and alarm functions.

In implementing mid-span meets, the network provider can no longer depend on a single vendor as a partner, but must mediate among multiple vendors. In single-vendor installations, responsibility for problems at installation, turn-up or during operation can be placed clearly with the equipment vendor. In multi-vendor environments, however, the network provider must have the tools and procedures necessary to identify which vendor's equipment is at fault should problems arise.

This requires an external test capability to serve as a "referee" and perform the mediation function. Each terminal in turn can be replaced by this known standard test interface. The test set then generates SONET signals into the terminal, and measures the terminal's output, verifying compliance and operation of the equipment. In this way, a failed network element can be isolated.

International Gateways: Check the H1 Byte!

- European (SDH) systems set bits 5&6 of the H1 pointer byte to 01.

- Some US equipment can't tolerate this value.

- New international trunks to the US often follow the SDH standard.

- Check the H1 byte value when experiencing difficulties with SONET equipment interfacing at these international gateways.

New international fiber installations are adopting SDH (Synchronous Digital Hierarchy) standards. While similar in most respects to SONET, the SDH signal asserts a 01 value in bits 5 and 6 of the H1 pointer byte. The SONET standard identifies these bits as "don't care." This means that SONET systems should be able to tolerate any value in these bits. In practice, however, SONET equipment may not accept these signals, resulting in alarm or failure conditions.

It is useful to check the H1 byte on signals at international gateways before connection to domestic SONET equipment.

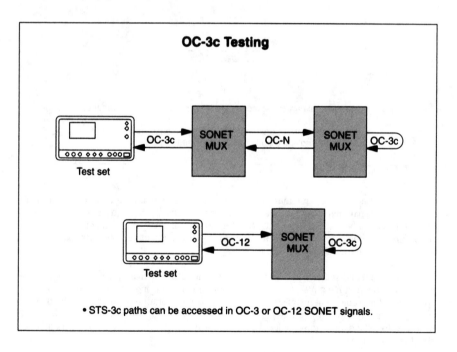

OC-3 transport of ATM signals makes use of the bandwidth of 3 STS-1 signals **concatenated** into a single STS-3c payload. Clear channel STS-3c testing is required to pre-qualify transmission links before ATM traffic is applied. The STS-3c payload is transmitted within either an OC-3 or higher-rate OC-N signal.

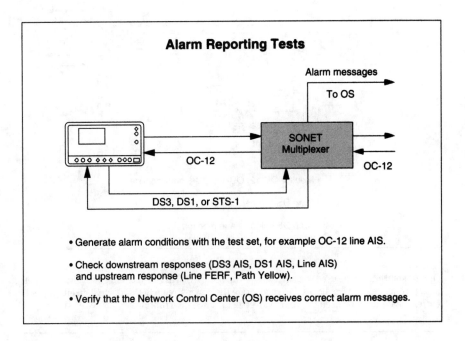

Alarm detection and processing can be tested through controlled external generation and detection of alarm conditions in the SONET and tributary signals.

For each category of SONET equipment, specific up-stream and down-stream alarm responses are required. On detection of an incoming alarm condition such as Loss of Signal (LOS), or Loss of Frame (LOF), the equipment must send appropriate indications toward the trouble signal (up-stream) and away from the trouble signal (down-stream). External test sets must provide the ability to force these alarm conditions, and to detect the alarm responses received from the unit under test.

In many cases, SONET terminals are connected to Operations Systems for reporting of alarm conditions. It is advantageous to verify that this alarm communication is working properly. Once again, the external test set's ability to force alarm conditions is used. The required alarm conditions can be forced, and the reports to the OS can be verified.

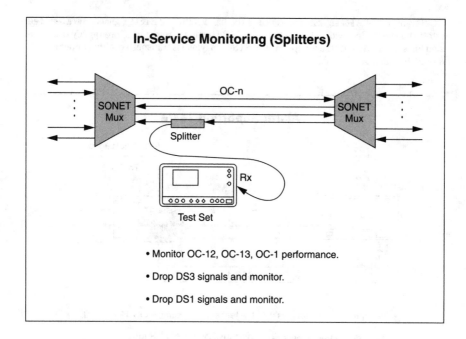

In-Service Monitoring (Splitters)

- Monitor OC-12, OC-13, OC-1 performance.
- Drop DS3 signals and monitor.
- Drop DS1 signals and monitor.

Optical splitters can be used to gain access to an optical SONET signal. Alternatively, a test set pass-through feature can be used to regenerate the down-stream SONET signal. In either case, extreme care must be used to avoid inadvertent service interruption.

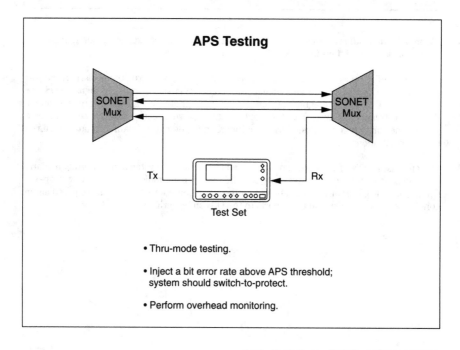

APS Testing

- Thru-mode testing.
- Inject a bit error rate above APS threshold; system should switch-to-protect.
- Perform overhead monitoring.

With a test set configured for through-mode testing, B2 Line Code Violations or line-level SONET alarms can be added to the signal forcing a switch to protection.

This tests point-to-point systems using line switching, and can also be applied to line-switched ring applications.

The Automatic Protection Switching is activated if line BIP error thresholds are exceeded. The error threshold levels are tested by injecting error rates slightly below and slightly above the nominal limit.

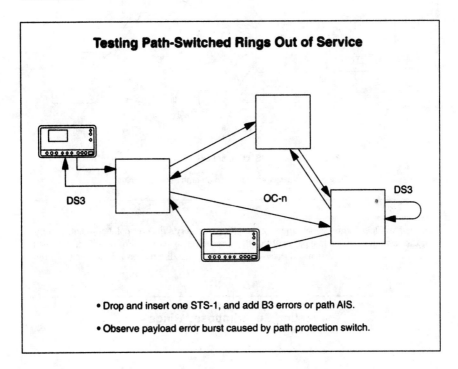

Testing Path-Switched Rings Out of Service

DS3

OC-n

DS3

- Drop and insert one STS-1, and add B3 errors or path AIS.
- Observe payload error burst caused by path protection switch.

In Unidirectional Path-Switched Rings, the decision to switch to protection is made independently for each path on the ring. This means that at any given time, certain paths may be using the main ring, while others are using the protection ring.

For out of service path-switched ring testing, a test set configured for SONET Drop & Insert can be placed in the optical line. B3 Path Code Violations or Path AIS alarm can be added to a selected STS-1 within the OC-3 or OC-12 signal. These violations should then be detected at the path terminating device, causing a path protection switch.

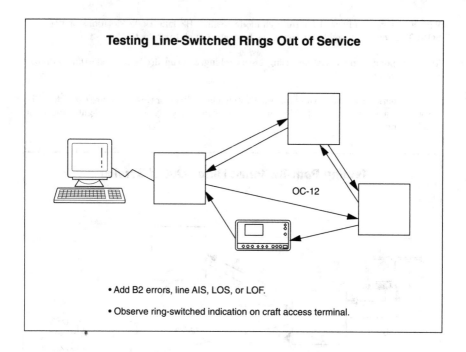

Testing Line-Switched Rings Out of Service

OC-12

• Add B2 errors, line AIS, LOS, or LOF.

• Observe ring-switched indication on craft access terminal.

Four or Two Fiber Bidirectional Line-switched Rings are very similar to point-to-point systems in that they switch based on line parameters (B2 errors and line alarms). For out-of-service testing, a test set in through mode can add these error and alarm conditions, forcing the ring to switch.

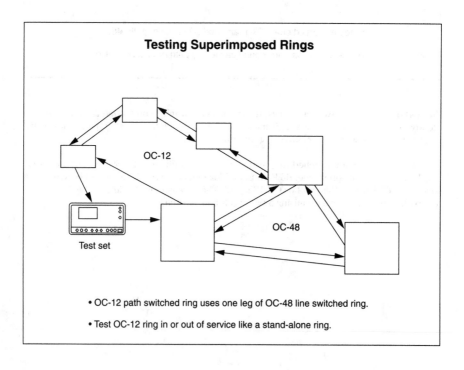

Testing Superimposed Rings

OC-12

OC-48

Test set

• OC-12 path switched ring uses one leg of OC-48 line switched ring.

• Test OC-12 ring in or out of service like a stand-alone ring.

Many networks are now employing multiple interconnected rings. For example, it is not uncommon for an OC-3 or OC-12 path switched ring to have one or more of its spans actually transported over a higher rate line-switched ring. In these cases, the lower rate ring can be accessed on one of its native spans just like a stand-alone path switched ring.

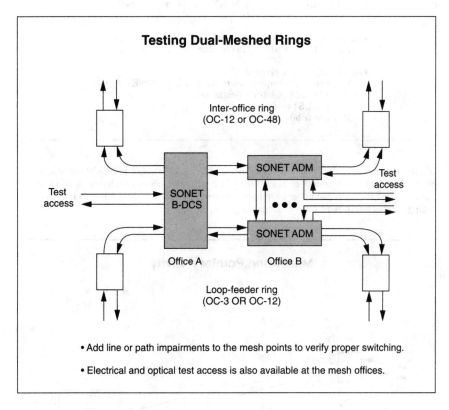

Testing Dual-Meshed Rings

Inter-office ring
(OC-12 or OC-48)

SONET ADM

Test access

Test access

SONET B-DCS

SONET ADM

Office A   Office B

Loop-feeder ring
(OC-3 OR OC-12)

• Add line or path impairments to the mesh points to verify proper switching.

• Electrical and optical test access is also available at the mesh offices.

Another ring inter-working configuration involves dual-meshed rings. Traffic from one ring to another is guaranteed even if one of the mesh offices goes out of service. This functionality can be tested by inserting impairments at the line or path level in one of the mesh interconnections. The traffic should switch to the other connection with no more than a protection error burst.

Ring mesh points also provide circuit access opportunities, through the test facilities of a Broadband Cross-connect System, or through low-speed ports on interconnected Add-Drop Multiplexers.

Timing and Synchronization

- Clock differences between SONET signals are compensated for by pointer adjustments.

- Each pointer adjustment results in a positive or negative "stuff" of 8 bits, resulting in tributary jitter.

- To identify timing problems:
  - Measure clock frequencies recovered from SONET signals.
  - Count SONET pointer adjustments.
  - Measure DS1 timing slips.
  - Measure tributary jitter.

Timing distribution is critical in SONET networks. In the installation and maintenance of SONET systems, it is therefore necessary to identify and correct faulty or drifting timing sources. Test methods can range from measurement of derived SONET clock signals to monitoring pointer adjustment rates or measuring tributary jitter.

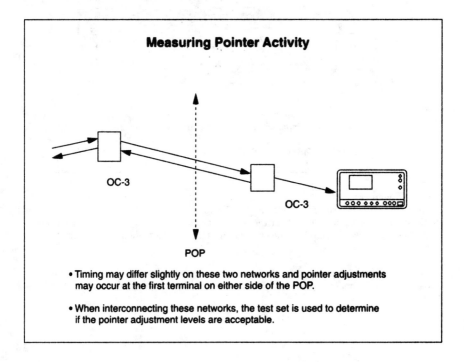

Measuring Pointer Activity

OC-3

OC-3

POP

- Timing may differ slightly on these two networks and pointer adjustments may occur at the first terminal on either side of the POP.

- When interconnecting these networks, the test set is used to determine if the pointer adjustment levels are acceptable.

The preceding illustration depicts an OC-3 connection between two networks at a Point-of-Presence. What is of greatest interest in connecting these two networks is the level of pointer activity caused by their interconnection. Since their clocks may be slightly different, pointer adjustments may occur at the first terminal on either side of the POP. In order to measure these adjustments, the OC-3 signal is monitored just *beyond* the first SONET multiplexer on either side of the network interconnection point.

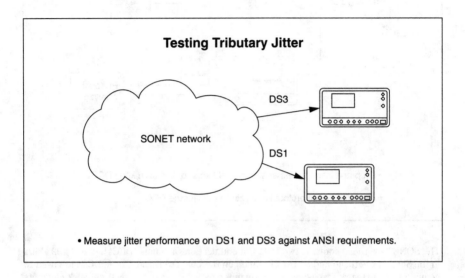

This makes tributary jitter measurement a must for networks carrying DS3 or DS1 services over SONET networks. Tributary jitter measurements allow the service provider to verify jitter performance against industry standard masks for DS1, E1, DS3, and STS-1 signals.

SONET pointer adjustments can cause jitter on tributary DS3 or DS1 signals. The jitter accumulates as the tributary signal crosses through multiple SONET islands, with potential impact on clock synchronization and error performance in the receiving terminals.

This makes tributary jitter measurement a must for networks carrying DS3 or DS1 services over SONET networks. Tributary jitter measurements allow the service provider to verify jitter performance against industry standard masks for DS1, E1, DS3, and STS-1 signals.

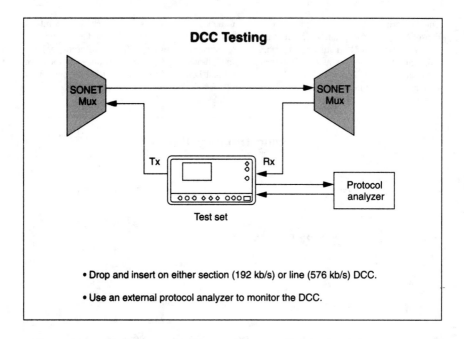

DCC Testing

- Drop and insert on either section (192 kb/s) or line (576 kb/s) DCC.

- Use an external protocol analyzer to monitor the DCC.

The SONET overhead contains two Data Communication Channels (DCCs), one at 192 kbit/s in the Section Overhead and one at 576 kbit/s on the Line Overhead. These DCCs are used to communicate network management and maintenance messages between network elements and the operations support computer system. The message-based information is transferred by means of an OSI protocol stack running at high speed.

The SONET Tester can provide access to either of the DCCs for an external Protocol Analyzer so that the test response messages may be isolated and analyzed.

Summary

- SONET installation and maintenance activity is increasing.

- Some SONET tests are similar to asynchronous methods.

- There are also new requirements like Alarm, APS, Ring and Pointer testing.

- The basic test scenarios discussed today can be applied in many SONET installation and maintenance test situations.

As SONET deployment continues, new test and maintenance issues will continue to emerge. SONET networks of the future will feature multiple rings, interfaces among SONET network elements at many different OC-N rates, and complex tributary mapping and routing plans. The ability to routinely access and test at these OC-N interfaces, and to map from any one interface rate to any other will be necessary for network implementers and maintenance personnel. And as SONET network complexity increases, so will the requirements for sophisticated test and diagnostic procedures.

SONET test challenges are evolving from compliance to installation and maintenance. An understanding of the changing installation and maintenance test requirements will assist in the rapid and efficient deployment of new revenue-generating SONET networks and services.

Glossary of Current Commonly Used Acronyms Relating to Telecommunications

ADC Analog-to-digital converter/conversion.

ANI Automatic number identification, a telephone industry term.

ANSI American National Standards Institute, a standard-setting organization.

ASIC Application-specific integrated circuit, a solid-state device.

ATE Automatic testing and evaluation.

ATM Asynchronous transfer mode, a type of data transfer switch.

BER Bit error rate, the number of bit errors per unit of time.

BERT Bit error rate tester/testing, an instrument that measures BER.

CDMA Code division multiple access, usually a LAN term. Also used by cellular telephone systems.

CDPD Cellular digital packet data, a cellular telephone term.

CD-ROM Compact disc read only memory, a computer term.

CISC Complex instruction set computer, a type of computer chip.

CSMA/CD Carrier sense multiple access/collision detection, a LAN term.

DAC Digital-to-analog converter/conversion, opposite of ADC.

DFT Design for testability, a design term for a system or product.

DFT Discrete Fourier transform, a mathematical term with reference to digital signal processing.

DRAM Dynamic random access memory, a computer memory circuit device.

DSP Digital signal processing, refers to voice, video, and computer-type digital signals.

EDI   Electronic data interchange, a LAN term.

e-mail   Electronic mail, a personal computer message system originating from the AT&T Easy Link System.

FDDI   Fiber Distributed Data Interface, an ANSI specification for a token-passing fiber-optic ring-based network.

FDMA   Frequency division multiple access, a LAN term.

FPGA   Field programmable gate arrays, a solid-state digital device.

FRAD   Frame relay access device, a LAN/WAN interconnect device.

FSK   Frequency shift keying, a carrier modulation method used by modems.

FSN   Full-service network, a selection of available services on a network, a term often used by cable television operators.

FTTC   Fiber to the curb, a cable television industry term.

FTTH   Fiber to the home, referring to a fiber-optic cable system.

FTTN   Fiber to the node, a term used by cable television and/or the telephone industry referring to fiber-optical plant.

GPIB   General-purpose instrument bus, instrument-to-computer interconnect.

ISDN   Integrated Services Digital Network, an all-transmission method for voice, video, and data services.

LAN   Local area network, a network for data communications.

LEC   Local exchange carrier, the local telephone system.

LLC   Logical link control, a LAN sublayer protocol.

MAC   Media access control, a LAN sublayer protocol.

MAP   Manufacturing automation protocol, a LAN/WAN term.

MPEG   Motion picture experts group, an organization that recommends compressed digital video signal standards (Standards I and II).

NIU   Network interface unit, a LAN device operating between the workstations and the network.

OSCA   Operating system's computing architecture, a LAN term originating from Bellcore.

OSI   Open system interface, a LAN term.

PBX   Private branch exchange, a telephone industry term for an in-building or corporate telephone system.

PCM   Pulse code modulation, a digital signal transmission method.

PCMIA   Personal computer memory card international association or personal computer multicard interconnect, a 68-pin connector for computer access.

PCN   Personal communications network, a mobile cellular telephone term.

PCS Personal communications service, a mobile cellular telephone term, same as above.

POTS Plain old telephone service, ordinary voice telephone services.

PSK Phase shift keying, a digital signal method of modulating a transmission carrier.

PSTN Public switched telephone network, a telephone industry term referring to the telephone system network.

RBOC Regional Bell operating company.

RISC Reduced instruction set computer, a microprocessor type of integrated circuit.

RS-232 Electronic Industry Association (EIA) interconnection specification between a computer workstation and a communication device.

SDH Synchronous digital hierarchy, a SONET term.

SMDS Switched multimegabit data service, a LAN/WAN interconnect term.

SNA System network architecture, a LAN/WAN term.

SNMP System network management protocol, a LAN network management term.

SONET Synchronous Optical Network, a fiber-optic digital transmission specification.

SRAM Static random access memory, a type of memory chip, a solid-state memory device.

TCP/IP Transmission Control Protocol/Interconnect Protocol. A LAN/WAN data transport layer.

TDMA Time division multiple access, a type of LAN system.

TOP Technical office protocol, a LAN term.

VISA Virtual instrument software architecture, a test instrument interconnect term.

VOD Video on demand, a cable television video service term.

WAN Wide area network, a data communication term.

Bibliography

Barker, Forrest. *Communications Electronics*. Prentice Hall, Englewood Cliffs, New Jersey, 1987.

Bartee, Thomas C., Editor in Chief. *Digital Communications*. Howard W. Sams and Co., Indianapolis, Indiana, 1986.

Bartlett, Eugene R. *Cable Television Technology and Operations: HDTV and NTSC Systems*. McGraw-Hill, Inc., New York, 1990.

Bartlett, Eugene R. *Electronic Measurements and Testing*, McGraw-Hill, Inc., New York, 1992.

Benson, K. Blair, Editor in Chief. *Television Engineering Handbook*. McGraw-Hill Inc., New York, 1986.

Buchsbaum, Walter H., *Buchsbaum's Complete Handbook of Practical Electronic Reference Data, Second Edition*. Prentice Hall, Englewood Cliffs, New Jersey 1978.

Dungan, Frank R. *Electrical Communication Systems*. Delmar Publishers, Inc. Albany, New York, 1993.

Fike, John L. PhD. Friend, George E. and Staff of The Texas Learning Center. *Understanding Telephone Electronics*, Radio Shack a Division of Tandy Corp., Ft. Worth, Texas, 1983.

Freeman, Roger L. *Telecommunications Transmission Handbook, Second Edition*. John Wiley & Sons, Inc., New York, 1981.

Heldman, Robert K. *Future Telecommunications*. McGraw-Hill Inc., New York, 1993.

Inglis, Andrew F. *Video Engineering*. McGraw-Hill Inc., New York, 1993.

Jordan, Edward C., Editor in Chief, *Reference Data For Engineers, Radio, Electronics, Computer and Communications, Seventh Edition*. Howard W. Sams & Co., Inc., Indianapolis, Indiana, 1985.

Lindberg, Bertil C. *Troubleshooting Communications Facilities*. John Wiley and Sons, Inc., New York, 1990.

Martin, James, Kathleen K. Chapman, The Arben Group, Inc. *Local Area Networks*. Prentice Hall, Englewood Cliffs, New Jersey, 1989.

Miller, Gary M. *Modern Electronic Communication*. Regents/Prentice Hall, Englewood Cliffs, New Jersey, 1993.

Peebles, Peyton Z. Jr. *Digital Communication Systems*. Prentice Hall, Inc., 1987.

Sterling, Donald J., Jr. *Technicians Guide to Fiber Optics*. Delmar Publishers, Inc., Albany, New York, 1993.

Winch, Robert G. *Telecommunication Transmission Systems*. McGraw-Hill, Inc., New York, 1993.

Suggested Periodicals and Magazines
Written for Telecommunications

CED, The Premier Magazine of Broadband Communications, P.O. Box 7699, Riverton, New Jersey 08077-7699. Tel. (609)786-0501, Fax (212)887-8493.

Communications Technology, Official Trade Journal of the SCTE, CT Publications Corp., 1900 Grant Street, Ste 720, Denver, CO 80803. Tel. (303)839-1565, Fax (303)839-1564.

Communications Week, The Newspaper for Enterprise Networking, CPM Publications, Inc., 600 Community Drive, Manhasset, New York 11030. For subscriptions, Circulation Dept., Communications Week, P.O. Box 1094, Skokie, Illinois 60076. Tel. (708)647-6834, Fax (708)647-6838.

EE, Evaluation Engineering, The Magazine of Electronic Evaluation & Test, Nelson Publishing, 2504 N. Tamiami Trail, Nokomis, FL   34275-3482. Tel. (813)966-9521, Fax (813)966-2597.

Electronic Design, Penton Publishing Inc., 1100 Superior Av, Cleveland, Ohio 44114-2543. Tel. (216) 696-7000. Subscription requests: Penton Publishing Subscription, Lockbox, P.O. Box 96732, Chicago, Illinois 60693.

IEEE Communications Magazine, IEEE Service Center, 445 Hoes Lane, Piscataway, New Jersey 08855-1331. Tel. (908)981-0060.

Lightwave, a Penwell Publication, Ten Tara Boulevard, Fifth Floor, Nashua, New Hampshire 03062-2801. For subscription inquiries only. Tel. (928)832-9349. Fax (918)832-9295.

Telecommunications, 685 Canton Street, Norwood, Massachusetts 02062. Tel. (617)769-9750, Circulation Tel. (617)356-4595, Fax (617)762-9071.

Via Satellite, Phillips Business Information, Inc., 1201 Seven Locks Road, Ste 300, Potomac, Maryland 20854. Tel. (301)340-1520, Fax (301)340-0542.

Index

ABOUT THE AUTHOR

Eugene R. Bartlett is a cable industry consultant and independent technical writer. He was formerly the director of engineering and partner in several cable television systems. Mr. Bartlett is the author of *Cable Television Technology and Operations* and *Electronic Measurements and Testing*, both published by McGraw-Hill.